LTE-A Cellular Networks

Abid Yahya

LTE-A Cellular Networks

Multi-hop Relay for Coverage, Capacity and Performance Enhancement

With contributions from

Jaafar A. Aldhaibani
R. Badlishah Ahmad
Joseph M. Chuma

 Springer

Abid Yahya
BIUST
Botswana International University of Science & Technology
Palapye, Botswana

ISBN 978-3-319-82784-1   ISBN 978-3-319-43304-2   (eBook)
DOI 10.1007/978-3-319-43304-2

Printed on acid-free paper

This Springer imprint is published by Springer Nature
The registered company is Springer International Publishing AG Switzerland
The registered company address is Gewerbestrasse 11, 6330 Cham, Switzerland

Dedicated to my family for their love, support, and sacrifice along the path of my academic pursuits

Abid Yahya

Dedicated to those exemplary personalities who strive for the betterment of mankind without personal gain

Jaafar A. Aldhaibani

Preface

Multi-hop relay is considered as one of the main keys for Long Term Evaluation—Advanced (LTE-A) to meet the growing demand for coverage extension and capacity enhancement. However, these benefits of multi-hop depend on the location of Relay Node (RN) which mitigates interference among the cells.

This book consists of five chapters and is organized as follows:

Chapter 1 gives an overview, clarifying the issues and motivating aspect of the research, together with the objectives, and overall book layout.

Chapter 2 provides a literature survey for various methods of relay deployment, and updates the state of current developments and solutions in the field of relay techniques, while evaluating the developments and solutions and critique of each method.

Chapter 3 gives a detailed explanation of three mathematical modeling techniques called Optimum RN Deployment (ORND), Enhance Relay Link Capacity (ERLC), and Balance Power Algorithm (BPA) within LTE-A cellular networks. ORND involves the mathematical derivation of the optimum RN location, an allocation of transmitted power for each RN, the optimum number of RNs within cell, the handover process, and the frequency reuse scheme. ERLC focuses on performance analysis by employing two antenna types in the RN to enhance relay link capacity. At the end of this chapter, the BPA is illustrated to minimize the transmission power consumption for MR.

Chapter 4 details the results from mathematical formulations and compares it with the simulation results in terms of spectral efficiency, coverage area, throughput, and transmission power consumptions for the MR using BPA.

Chapter 5 presents optimum location for relay node in LTE-A. This chapter concludes the originality and innovations with a summary of the results.

Palapye, Botswana

Abid Yahya
Jaafar A. Aldhaibani
R. Badlishah Ahmad
Joseph M. Chuma

Acknowledgments

Thank you to the following individuals without whose contributions and support this book would not have been written:

Authors would like to express their special gratitude and thanks to Botswana International University of Science and Technology (BIUST), Universiti Malaysia Perlis (UniMAP), Regent University College of Science and Technology (RUCT) Ghana, and University College of Science and Technology (RUCST) for giving us such attention, time, and opportunity to publish this book.

Authors would also like to take this opportunity to express their gratitude to all those people who have provided us with invaluable help in the writing and publication of this book.

Contents

Abbreviations

1G	First Generation
2G	Second Generation
3G	Third Generation
3GPP	Third Generation Partnership Project
3GPP-LTE	Third Generation Partnership Project-Long Term Evaluation
4G	Fourth Generation
16-QAM	16-Quadrature Amplitude Modulation
AF	Amplify and Foreword
AMC	Adaptive Modulation and Coding
AMPS	Advanced Mobile Phone Systems
AWGN	Additive White Gaussian Noise
BPA	Balance Power Algorithm
BPSK	Binary Phase-Shift Keying
BS	Base Station
CA	Carrier Aggregation
CDMA	Code Division Multiple Access
CoMP	Coordinated Multi-Point transmission and reception
CQI	Channel Quality Indicator
DA	Directional Antenna
DAS	Distributed Antenna System
DF	Decode and Foreword
DL	Downlink
eNB	Evolved Node B
ERLC	Enhance Relay Link Capacity
EPs	Extension Points
EVM	Error Vector Magnitude
FD	Full-Duplex
GSM	Global System for Mobile Communications
HD	Half-Duplex
ICI	Inter-cell Interference

IMT-A	International Mobile Telecommunications-Advanced
IMT-2000	International Mobile Telecommunications-2000
IP	Internet Protocol
ITU	International Telecommunication Union
LOS	Line-of-Sight
LTE	Long Term Evaluation
LTE-A	Long Term Evaluation—Advanced
MANETs	Mobile Ad Hoc Networks
MCS	Modulation and Coding Scheme
MIMO	Multiple-Input Multiple-Output
MR	Moving Relay
NLOS	Non-Line-of-Sight
OA	Omni-directional Antenna
OFDM	Orthogonal Frequency Division Multiplexing
OFDMA-TDD	Orthogonal Frequency Division Multiple Access-Time Division Duplexing
ORND	Optimum Relay Node Deployment
QoS	Quality of Service
QPSK	Quadrature-Phase-Shift Keying
RN	Relay Node
RSS	Received Signal Strength
SER	Sample Error Rate
SNR	Signal to Noise Ratio
SINR	Signal to Interference-Plus-Noise Ratio
TDMA	Time Division Multiple Access
UE	User Equipment
UMTS	Universal Mobile Telecommunications System
UL	Uplink
VPL	Vehicle Penetration Loss
WiMAX	Worldwide Interoperability for Microwave Access
WLAN	Wireless Local Area Network

List of Symbols

BW_{eff}	Adjustment for bandwidth efficiency
$BW_{\phi az}$	Beam width pattern at azimuth angle
$BW_{\theta el}$	Beam width pattern at elevation angle
C_{max}	Upper limit spectral efficiency for BS
$C_{R\,max}$	Upper limit spectral efficiency for RN
$C_{RN,2}$	Upper limit spectral efficiency for RN at Location 2
$C_{RN,3}$	Upper limit spectral efficiency for RN at Location 3
C_{sim}	Spectral efficiency for BS through simulator
C_i	Estimated spectral efficiency for BS
d	Distance between BS and UE
d_{nr}	Distance between neighboring RNs
$d_{nr,j}$	Distance between RNs in neighboring cell
d_A	Distance between BS and MR above vehicle
$d_{c,q}$	Distance between BS and UE inside vehicle
$d_{i,k}$	Distances between UE and BS$_j$
D_{RN}	Location of RN from BS
$D_{i,k}$	Distances between UE to BS$_i$
$E\,[.]$	Expectation function
G_t	Antenna gains for transmitter
G_{re}	Antenna gain of receiver
G_{tr}	Antenna gain of transmitter
G_d	Gain of DA for RN
G_{ue}	Antenna gain for UE
G_{BS}	Antenna gain for BS
G_r	Antenna gains for receiver
$H_{i,k}$	Fading channel gain form donor to user
$H_{j,k}$	Fading channel gain from neighboring cell to user
$H_{i,xs}$	Fading channel gain form donor BS to user at X_s location
$H_{i,2}$	Fading channel gain form donor BS to user at location 2
$H_{i,3}$	Fading channel gain form donor BS to user at location 3

$H_{i,xo}$	Fading channel gain form donor BS to user at X_o location
$H_{RN,xo}$	Fading channel gain form RN to user at X_o location
$H_{j,xs}$	Fading channel gain from neighboring cell to user at X_s location
H_A	Fading channel at relay link
H_B	Fading channel at access link
H_C	Fading channel at direct link
$H_{k,q}$	Matrix fading channel between kth-RN and qth-user
L_r	RN characteristics
L_{re}	Feeder losses at receiver
L_t	Feeder losses at transmitter
L_{prop}	Propagation loss
L_{fsd}	Free space distance loss
L_d	Diffraction loss
L_{sp}	Sub-path loss
L_{gas}	Attenuation caused by atmospheric gas
L_{rain}	Attenuation caused by hydrometeor scatter
L_{clut}	Clutter attenuation
N_k	Background noise at user
N_{Xo}	Background noise for user at X_o
N_{xs}	Background noise for user at X_s
N_2	Background noise at user in location 2
N_3	Background noise at user in location 3
N_{relays}	Optimum number of relay
N_{cell}	Number of neighboring cell
P_i	Transmitted power from BS
P_{UE}	Transmitted power from UE
P_{RN}	Transmitted power from RN
P_{rj}	Received power to UE from neighboring BS
P_j	Transmitted power from neighboring BS
$P_{o,RL}$	Outage probability of relay link
$P_{o,MH}$	Outage probability of multi-hop link
$P_{o,access}$	Outage probability of access link
$\rho_{RN,Xo}$	SINR for user at Xo via RN
$\rho_{i,k}$	SINR at k-user via BS_i
ρ_i	SINR for each user in the cell
ρ_{ideal}	Ideal SINR for user at X_s location
ρ_{max}	Maximum limitation on received SINR by using EVM
$\rho_{i,xs}$	SINR from BS to user at X_s location
$\rho_{RN,2}$	SINR from RN to user at Location 2
$\rho_{RN,3}$	SINR from RN to user at location 3
$\rho_{UE,q}$	Downlink SNR at user via direct and relay links
ρ_{BS}	Uplink SNR at BS via direct and relay links
$\rho_{UE,q}^{gm}$	SNR at UE inside vehicle (Group Mobility) via MR
$\rho_{UE,q}^{Direct}$	SNR at UE inside vehicle (group mobility) via direct link

ρ^{max}	Maximum required of SNR at UE inside vehicle
ρ_{th}	Threshold of SNR at UE inside vehicle
ρ_{eff}	Adjustment for SINR spectral efficiency
p_{t}	Transmitted power from source
p_{r}	Received power from destination
$p_{\mathrm{multi\text{-}hop}}^{\mathrm{r}}$	Received power via multi-hop link
$p_{\mathrm{traditional}}^{\mathrm{r}}$	Received power via traditional link
$R_{\mathrm{UE},q}^{\mathrm{gm}}$	Bit Rate at UE inside vehicle (Group Mobility) via MR
$R_{\mathrm{UE},q}$	Downlink bit rate at user via direct and relay links
R	Cell radius
$\mathrm{SINR}_{\mathrm{sim}}$	SINR through simulator
T_{MR}	Time of approaching of vehicle
V_{MR}	Velocity of vehicle
$X_{i,k}$	Received signal from BS for user
$X_{j,k}$	Received signal from neighboring BS for user
$X_{i,xs}$	Received signal from BS for user at X_s location
$X_{i,2}$	Received signal from BS for user at location 2
$X_{i,3}$	Received signal from BS for user at location 3
$X_{\mathrm{RN},2}$	Received signal from RN for user at location 2
$X_{\mathrm{RN},3}$	Received signal from RN for user at location 3
$X_{j,xs}$	Received signal from neighboring BS at user at location X_3
$X_{\mathrm{RN}}[t_2]$	Transmitted signal form RN at Second time slot $[t_2]$
X_{s}	Distance of estimated saturation capacity
$X_{\mathrm{s}2}$	Distance of estimated saturation capacity for RN at location 2
$X_{\mathrm{s}3}$	Distance of estimated saturation capacity for RN at location 3
$Y_{i,k}$	Received signal for k-user from BS_i
$Y_{\mathrm{RN},Xo}$	Received signal for user from RN at Xo location
$Y_{\mathrm{RN},2}$	Received signal for user from RN at location 2
$Y_{\mathrm{RN},3}$	Received signal for user from RN at location 3
Y_{RL}	Received signal for user via relay link
Y_{AL}	Received signal for user via access link
$y_{\mathrm{UE},q}[t_1]$	Received signal for user at first time slot $[t_1]$
$y_{\mathrm{UE},q}[t_2]$	Uplink received signal by UE_q at second time slot $[t_2]$
$y_{\mathrm{RN}}[t_1]$	Received signal for RN at first time slot $[t_1]$
$y_{\mathrm{BS}}[t_2]$	Downlink received signal by BS at second time slot $[t_2]$
$\hat{y}_{\mathrm{BS}}[t_2]$	Downlink received signal by BS at second time slot $[t_2]$ after cancelation of the self-interface
$\hat{y}_{\mathrm{UEq}}[t_2]$	Uplink received signal by UE_q second time slot $[t_2]$ after cancelation of the self-interface
γ_{th}	Certain threshold of SINR
γ_{RL}	SINR at the relay link
γ_{access}	SINR at the access link
α	Path loss exponent
λ	Wavelength of the carrier frequency

θ_{el}	Elevation angle for antenna
θ_{az}	Azimuth angle for antenna
X_o	Handover distance from BS
σ_o	Variance
Ψ	Amplification factor for AF relay

About the Authors

Prof.Abid Yahya is an esteemed scientist who graduated with a B.Sc. degree from the University of Engineering and Technology, Peshawar, Pakistan, in Electrical and Electronic Engineering, majoring in telecommunication. Prof. Yahya began his career on a path that is rare among other research executives and earned his M.Sc. and Ph.D. degree in Wireless and Mobile systems in 2007 and 2010, respectively, from the Universiti Sains Malaysia, Malaysia. He has over 100 research publications to his credit in various books, research journals of repute, and conference proceedings. Prof. Abid Yahya has supervised a number of Ph.D. candidates.

Dr. Jaafar A. Aldhaibani earned his B.Sc. degree in Electrical and Electronic Engineering, major in Telecommunications, from the University of Technology, Baghdad, Iraq and earned his PhD and M.Sc. degree in Wireless Communications from UniMap and the University of Technology, Baghdad, Iraq respectively. Dr. Jaafar worked with the Motorola Company (AIEE) and ATDI Company for RF planning communication sites. Currently he is a head of the Networks Communication Research Department in Ministry of Higher Education and Scientific Research, Iraq.

Prof. R. Badlishah Ahmad obtained his B.E. in Electrical and Electronic Engineering from Glasgow University in 1994. He received his M.Sc. and Ph.D. in 1995 and 2000, respectively, from the University of Strathclyde, UK. His research interests are computer and telecommunication network modeling using discrete event simulators, optical networking, and coding and embedded system based on GNU/Linux for vision. He has five years' teaching experience in Universiti Sains Malaysia. Since 2004 he has been working with Universiti Malaysia Perlis (UniMAP) as the Dean at the School of Computer and Communication Engineering.

Prof. Joseph M. Chuma received a B.E. in Electrical and Electronic Engineering from the University of Nottingham, UK, in 1992; an M.Sc. in Telecommunications and Information Systems; and a Ph.D. in Electronic Systems Engineering from the University of Essex, UK, in 1995 and 2001, respectively. His main areas of research are the design of compact single and dual-mode dielectric resonator filters for mobile communications. Prof Chuma has served as the Dean of the Faculty of Engineering and Technology at the University of Botswana. He is also serving as a Board Member of a parastatal organization in Botswana.

Chapter 1
Introduction to LTE Cellular Networks

1.1 Background

Mobile communications have come a long way since the introduction of the first mobile telephone systems in the 1950s by German National Railway (Dahlman, Parkvall, & Skold, 2011). Aside from the fact that the user equipments (UEs) were bulky and power hungry, there were other serious issues such as frequency allocation and the transmitted power of the base station (BS). Therefore, these primitive services severely limited the number of active users to the number of channels assigned to a particular frequency zone (Sesia, Toufik, & Baker, 2011).

In order to support advanced services and applications, the 3GPP throughout 2009 worked on a study with the purpose of identifying the LTE improvements required to meet IMT-Advanced requirements (Akyildiz, Gutierrez-Estevez, & Reyes, 2010). On September 2009, the 3GPP made a formal submission to the ITU so that LTE-Advanced could become candidate for IMT-Advanced.

This book presents a new model to enhance both capacity and coverage area in LTE-A cellular network by determining the optimum location for the RN with limited interference (Aldhaibani, Yahya, & Ahmad, 2012a; Aldhaibani, Yahya, Ahmad, Md Zain, & Salman, 2013; Aldhaibani, Yahya, Ahmad, Fayadh, & Abbas, 2014). A new model is designed to enhance the capacity of the relay link by employing two antennas in RN. This design enables the relay link to absorb more users at cell edge regions. An algorithm called the Balance Power Algorithm (BPA) is developed to reduce MR power consumption.

One of the solutions to meet growing demand and stringent design requirements for coverage extension, throughput, and capacity enhancement in LTE cellular networks is to increase the number of BSs (Due to the assumption of focusing this book on LTE system), refer to a base station (BS) by the 3GPP-LTE and sometimes called Node B (eNB) (3GPP, TS. ETSI, 2007), so that each station covers a small area (Meko, 2012; Yongchul & Sichitiu, 2011). However, increasing the number of BSs requires high deployment cost, increases interference between

© Springer International Publishing Switzerland 2017
A. Yahya, *LTE-A Cellular Networks*, DOI 10.1007/978-3-319-43304-2_1

stations, and requires extra spectrum (Chandrasekhar, Andrews, & Gatherer, 2008; Khakurel, Mehta, & Karandikar, 2012; Liu, Hoshyar, Yang, & Tafazolli, 2006). Hence, a cost-effective solution is needed to cover the required area while providing acceptable SINR to subscribers of LTE cellular networks, as well as to meet exponential demand in terms of coverage and capacity in future cellular networks with lower resource consumption (Dahlman et al., 2011). The term "resource" here refers to the frequency and power which is allocated by the network operators.

Cell edge region is a region in which the users suffer from outage in wireless services due to the poor coverage; in general, it is limited from 0.6R to R where R is coverage cell radius (Dahlman et al., 2011). RN is an appropriate solution to address low SINR at the cell edge region, resolve coverage holes due to shadowing and Non-Line-Of-Sight (NLOS) connections, and to meet the access requirement of nonuniform distributed traffic in densely populated areas in order to improve coverage and capacity (Pabst, Esseling, & Walke, 2005; Yang et al., 2009).

This book demonstrates a new way to improve the capacity, coverage, and optimization in power allocation of LTE-A cellular network (Aldhaibani, Yahya, Ahmad, & Md Zain 2012b Aldhaibani, Yahya, Ahmad, & Md Zain, 2013; Aldhaibani, Yahya, Ahmed, Ali, & Fayadh, 2014). It involves the design and development of new methods of multi-hop relay deployment with limited interference.

In this book, three different methods are presented to enhance the capacity and coverage area in LTE-A cellular networks. The scope involves the evaluation of the effect of the RN location in terms of capacity and the determination of the optimum location of the relay that provides maximum achievable data rate for users with limited interference at the cell boundaries. This book addresses frequency reuse by exploiting the available spectrum for RNs and alleviates interference among stations (BSs and RNs). The second method demonstrates a new design to improve relay link capacity to accommodate growing numbers of users at the cell edge region. Finally, the third method focuses on reducing the transmission power consumption for the MR using the Balance Power Algorithm.

In this book, three different methods to enhance the capacity and converge area for LTE cellular network are proposed, and the contribution of this book can be summarized as follows:

- The first method, ORND, is based on a mathematical analysis of modified Shannon formula of capacity distribution that provides a realistic downlink transmission compared to classical Shannon expressions as presented in Appen dix A. ORND provides mathematical formulations for saturation capacity distance (X_s), handover distance (X_o), optimal location for the RN (D_{RN}), and an optimum number of relays (N_{relays}) per cell. ORND focuses on the mitigation of interference and alleviation overlapping among the cells by balance of power transmission and frequency assignment.
- The second method, ERLC, is a new design, which enhances the capacity of the relay link by employing two types of antenna for the RN; Omni directional antenna (OA) provides the link with users, and directional antenna (DA) carries

information for users attached with the RN to the BS. ERLC enables the relay link to accommodate the growing number of users at cell edge regions and reduces the probability of outage.

- The third method, BPA, is a new algorithm and is installed in the MR to minimize transmission power consumption for the MR as well as provide a higher throughput for public transportation users. BPA is based on the performance evaluation of DL and UL transmission for the multi-hop mobility system.

References

3GPP, TS. ETSI (2007). *Base Station (BS) radio transmission and reception* (Version 8.0.0). 3GPP TS 36.104 [Technical Specification]. Release 8, Valbonne, France.

Akyildiz, I. F., Gutierrez-Estevez, D. M., & Reyes, E. C. (2010). The evolution to 4G cellular systems: LTE-advanced. *Physical Communication, 3*(4), 217–244.

Aldhaibani, J. A., Ahmad, R. B., Yahya, A., Azeez, S. A., & Abbas, A. H. (2014). Performance analysis of amplify and forward relay during uplink and downlink in LTE-A cellular networks. *Journal of Next Generation Information Technology, 5*(1), 1–8.

Aldhaibani, J. A., Yahya, A., Ahmad, R. B., Fayadh, R. A., & Abbas, A. H. (2014). Reducing transmitted power of moving relay node in LTE-A cellular networks. *Journal of Computer Science, 10*(6), 1051–1061.

Aldhaibani, J. A., Yahya, A., Ahmad, R. B., Md Zain, A. S., & Salman, M. K. (2013). On coverage analysis for LTE-A cellular networks. *International Journal of Engineering and Technology, 5*(1), 492–497.

Aldhaibani, J. A., Yahya, A., Ahmad, R. B., & Md Zain, A. S. (2012a). Performance enhancement of LTE-A, a multi-hop relay node, by employing half-duplex mode. *International Journal of Computer Science Issues, 9*(3), 273–280.

Aldhaibani, J. A., Yahya, A., Ahmad, R. B., & Md Zain, A. S. (2012b). Effect of relay location on two-way DF and AF relay in LTE-A cellular networks. *International Journal of Electronics and Communication Engineering & Technology, 3*(2), 385–399.

Aldhaibani, J. A., Yahya, A., Ahmad, R. B., & Md Zain, A. S. (2013). Performance analysis of two-way multi-user with balance transmitted power of relay in LTE-A cellular networks. *Journal of Theoretical and Applied Information Technology, 51*(1), 183–190.

Aldhaibani, J.A., Yahya, A., Ahmed, R.B., Ali, Z. G., & Fayadh, R.A. (2014). Enhancing link quality in a multi-hop relay in LTE-A employing directional antenna. In *2013 I.E. International RF and Microwave Conference*, Penang, 9–11 December 2013.

Aldhaibani, J.A., Yahya, A., Ahmed, R.B., Omar, N., & Ali, Z.G. (2013). Effect of relay location on two-way DF and AF relay for milt-users system in LTE-A cellular networks. In *Business Engineering and Industrial Applications Colloquium (BEIAC), IEEE International Conference*, 7–9 April 2013, Langkawi. New York: Institute of Electrical and Electronics Engineers.

Aldhaibani, J.A., Yahya, A., Ahmed, R.B., & Azeez, S.A. (2014). Increasing the coverage area through relay node deployment in long term evolution advanced (LTE-A) cellular networks. In *International Conference on Mathematics, Engineering & Industrial Applications 2014*, 28–30 May 2014. Melville, NY: American Institute of Physics (AIP).

Aldhaibani, J. A., Yahya, A., & Ahmad, R. B. (2012). Improvement of relay link capacity in a multi-hop system by using a directional antenna in LTE-A cellular network. *Przegląd Elektrotechniczny, 89*(11), 195–201.

Aldhaibani, J.A., Yahya, A., & Ahmed, R.B. (2014). Performance enhancement of moving relay node in LTE-A cellular networks. In *International Postgraduate Conference on Engineering and Management 2014 (Indexed by UniMAP)*.

Chandrasekhar, V., Andrews, J., & Gatherer, A. (2008). Femtocell networks: A survey. *Communications Magazine, IEEE, 46*(9), 59–67.

Dahlman, E., Parkvall, S., & Skold, J. (2011). *4G: LTE/LTE-advanced for mobile broadband: LTE/LTE-advanced for mobile broadband.* Chichester: Academic Press.

Aldhaibani, J. A., Yahya, A., & Ahmad, R. B. (2014). Coverage extension and balancing the transmitted power of the moving relay node at LTE-A cellular network. *The Scientific World Journal, 2014*(815720), 1–10.

Khakurel, S., Mehta, M., & Karandikar, A. (2012). *Optimal relay placement for coverage extension in LTE-A cellular systems.* Paper presented at the 2012 National Conference on Communications (NCC), Kharagpur.

Liu, Y., Hoshyar, R., Yang, X., & Tafazolli, R. (2006). Integrated radio resource allocation for multihop cellular networks with fixed relay stations. *IEEE Journal on Selected Areas in Communications, 24*(11), 2137–2146.

Meko, S. F. (2012). Optimal relay placement schemes in OFDMA cellular networks. *International Journal of Engineering Research and Applications, 2*(4), 1501–1509.

Pabst, R., Esseling, N., & Walke, B. H. (2005). Fixed relays for next generation wireless systems—System concept and performance evaluation. *Journal of Communications and Networks, 7*(2), 104–114.

Sesia, S., Toufik, I., & Baker, M. (2011). *LTE: The UMTS long term evolution.* Chichester: Wiley.

Yang, K., Hu, H., Xu, J., & Mao, G. (2009). Relay technologies for WiMAX and LTE-advanced mobile systems. *Communications Magazine, IEEE, 47*(10), 100–105.

Yongchul, K., & Sichitiu, M. L. (2011). Optimal placement of transparent relay stations in 802.16 j mobile multihop relay networks. *IEICE Transactions on Communications, 94*(9), 2582–2591.

Chapter 2
Opportunities, Challenges, and Terms Related to LTE-A Cellular Network

2.1 Introduction

The well-known generational phase of mobile telecommunications started in the early 1980s, beginning with the introduction of the so-called First Generation (1G) of mobile telecommunications standards. The 1G analog cellular systems supported voice communication with limited roaming and short-range radio waves telephones. Later, Second Generation (2G) was introduced as a digital systems and promised higher capacity and better voice quality. 2G cellular telecom networks were commercially launched on GSM (Global Systems for Mobile Communication) standard in Finland in 1991 (Dahlman, Parkvall, & Skold, 2011a).

The 3GPP (Third Generation Partnership Project) was born out of the International Telecommunication Union's (ITU) of International Mobile Telecommunications "IMT-2000" initiative, covering high speed, broadband, and Internet Protocol (IP) (Khan, 2009).

LTE (Long Term Evaluation) standardization within the 3GPP has reached a mature state since end of 2009. LTE mobile communication systems have been deployed as a natural evolution of GSM and UMTS (Universal Mobile Telecommunications System).

The ITU has coined term IMT-Advanced to identify mobile systems whose capabilities go beyond those of IMT-2000 (Dahlman, Parkvall, & Skold, 2011b; Khan, 2009). Figure 2.1 shows the cellular network generations of standards for mobiles communication and Table 2.1 shows summary of 3GPP-LTE, LTE-Advanced, and IMT-Advanced performance targets.

The rest of this chapter is organized as follows: A survey of basic concepts of using relaying technologies and advantages and disadvantages within cellular network is presented in Sects. 2.3 and 2.4. Section 2.5 introduces the types of relays based on their functionality. A literature survey of relay nodes related with cellular network is presented in Sect. 2.6. A relay enhance cellular network is reviewed in Sect. 2.7. Section 2.8 introduces types of relay transmission mode in the cellular

© Springer International Publishing Switzerland 2017
A. Yahya, *LTE-A Cellular Networks*, DOI 10.1007/978-3-319-43304-2_2

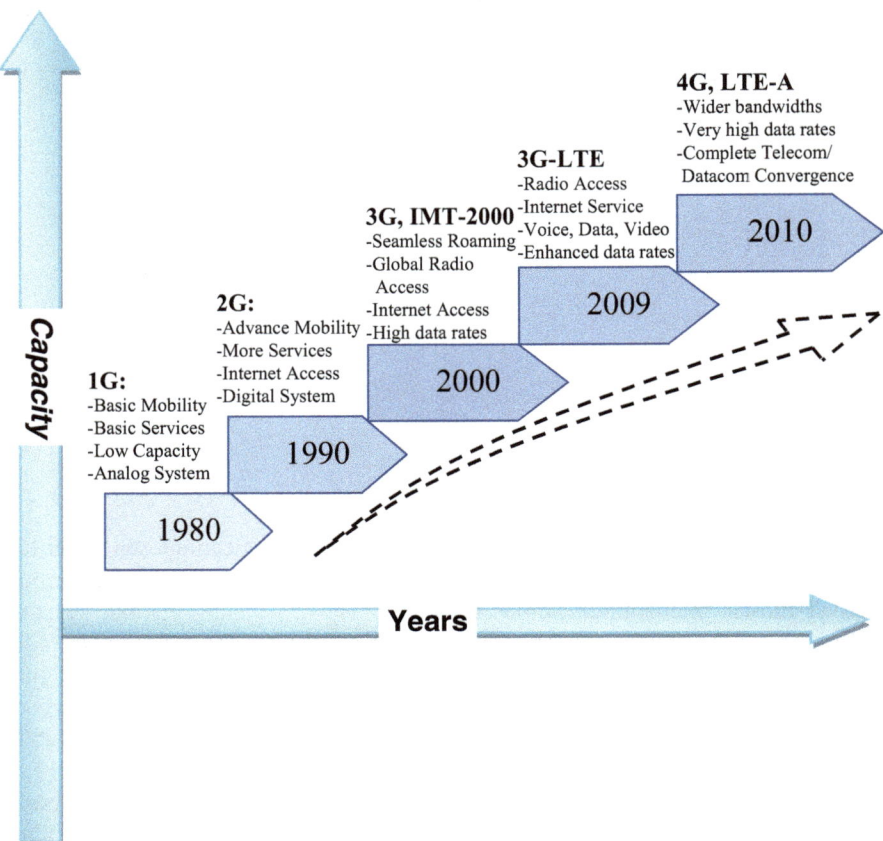

Fig. 2.1 Cellular network generations for standard mobiles communication (Dahlman et al., 2011a)

Table 2.1 3GPP-LTE, LTE-advanced, and IMT-advanced performance targets for downlink (DL) and uplink (UL) (Akyildiz et al., 2010)

Parameters	3GPP-LTE	LTE-Advanced	IMT-Advanced
Peak data rate (DL) (Mbps) (UL)	100	1000	1000
	50	500	1000
Spectral efficiency (DL) (bps/Hz) (UL)	5	30	15
	2.5	15	6.75

network. Section 2.9 is dedicated to relay planning in cellular network and shows the significance of relay location on increasing coverage area and enhancing throughput of the LTE cellular network. A new type of relay known as the moving relay is reviewed in Sect. 2.10. Finally, a summary of this chapter is presented in Sect. 2.11.

2.2 Long Term Evolution Advanced (LTE-A)

Long Term Evolution-Advanced (LTE-A) is an evolution of 3GPP-LTE (Third Generation Partnership Project-Long Term Evolution) which aims to bridge the gap between Third Generation (3G) and Fourth Generation (4G) standards described in IMT-Advanced (International Mobile Telecommunications). LTE-A aims to provide peak data rates of up to 1 Gbps (for low mobility) and 100 Mbit/s (for high mobility) in Downlink (DL) and 500 Mbps in Uplink (UL). LTE-A is required to reduce the latency time as compared to 3GPP-LTE (Dahlman et al., 2011a). LTE-Advanced targets to enhance the cell edge user throughput in order to achieve a homogeneous user experience in the cell and increase the capacity to 30 and 15 bps/Hz in DL and UL, respectively (Sesia, Toufik, & Baker, 2011).

The main new functionalities introduced by LTE-A to enhance 3GPP-LTE are carrier aggregation (CA), enhanced use of multi-input multi-output (MIMO) antenna techniques, coordinated multipoint transmission and reception (CoMP), and support by relay technology (Khan, 2009), as shown in Fig. 2.2.

- *Carrier Aggregation*: In carrier aggregation, multiple carrier components are aggregated, to provide wider bandwidths for transmission purposes both in DL and UL. It allows for transmission bandwidths of up to 100 MHz, by adding five component carriers of 20 MHz bandwidth. CA exploits the fragmented spectrum by aggregating contiguous or noncontiguous component carriers (Dahlman et al., 2011b).
- *Extended MIMO*: LTE-A introduced extending the number of layers in MIMO from 4×4 to 8×8 layer at DL and from 2×2 to 4×4 layers at UL to increase the overall bit rate through transmission of different data streams in multiple antennas (Sesia et al., 2011; Ullah, 2012).
- *Coordinated Multipoint Transmission/Reception (CoMP)*: In CoMP multiple geographically separated base station sites coordinate transmission and reception, in order to achieve good system performance and end-user service quality. CoMP uses coordination techniques; namely, intercell scheduling coordination and joint transmission/reception (Dahlman et al., 2011b; Genc, 2010).

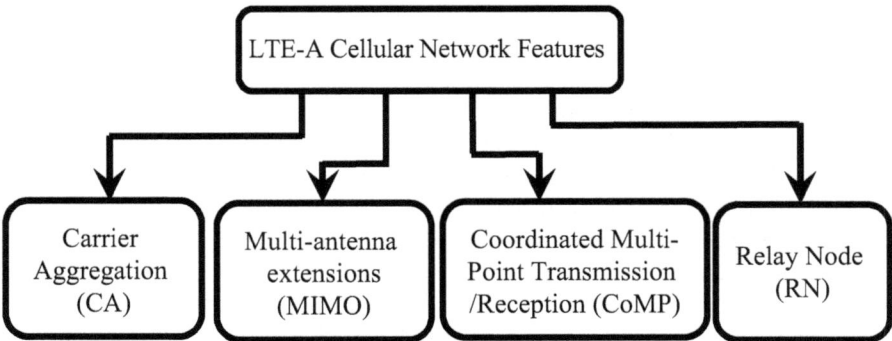

Fig. 2.2 Functionalities introduced by LTE-A to enhance 3GPP-LTE network

– *Relaying Technology*: This technique provides the possibility for heterogeneous network planning through the integration of large cells such as BS, and small cells such as Relay Node (RN). RNs have lower power and a lower cost compared to BSs which provide enhanced coverage and capacity in cellular networks (Dahlman et al., 2011a).

Relaying technologies enable the efficient utilization of communication resources, by permitting the nodes to cooperate in information exchange with each other to enhance the quality of services for current communication networks to suit LTE-A requirements.

This book focuses on exploiting the possibilities of RN to enhance the coverage and capacity for LTE-A cellular networks. This chapter presents surveys of related literature about relaying to enhance 3GPP-LTE cellular networks; Fig. 2.3 shows the chapter scope for the literature survey.

2.3   Cooperative Relaying

Cooperative relaying technologies have been presented as the basic concept of the relaying through proposing three terminals: a source, relay, and destination as shown in Fig. 2.4. The source broadcasts a signal to both the relay and the destination, while the relay rebroadcasts this signal to the destination. Thereafter, the destination combines these received signals from both the source and the relay to improve reliability (Duong & Zepernick, 2009).

Cooperative relaying is a technique for wireless communications, which promises gains in throughput and capacity and allows communication terminals in a network to receive and help the information transmission from each other. Cooperative relaying is based on exploiting the advantage of the broadcasting nature of wireless communications (Sharma, Shi, Hou, Sherali, & Kompella, 2010). Cooperative relaying for wireless communications has become one of the areas of active researches for more than three decades (Nordio, Chiasserini, & ElBatt, 2012).

The classical relay channel model, which consists of a source, a relay, and a destination, was first introduced by Van Der Meulen (1971). He discovered upper and lower bounds on the capacity of the relay channel and provided many observations which led to an improvement of his results. In the following years, Cover and Gamal (1979) significantly improved on Meulen's observations when they established an achievable lower bound to the capacity of the general relay channel. However, the implicit assumptions (Cover & Gamal, 1979; Van Der Meulen, 1971) are unrealistic for the wireless medium because they are not compatible with the current developments in communications systems (Yongchul & Sichitiu, 2011).

In routing scenario, cooperative relaying has two categories, Ad Hoc networks and multihop relay (Sharma & Jain, 2010) which use an independent node known as Relay Node (RN) or sometimes known as Relay Station (RS). In Ad Hoc network

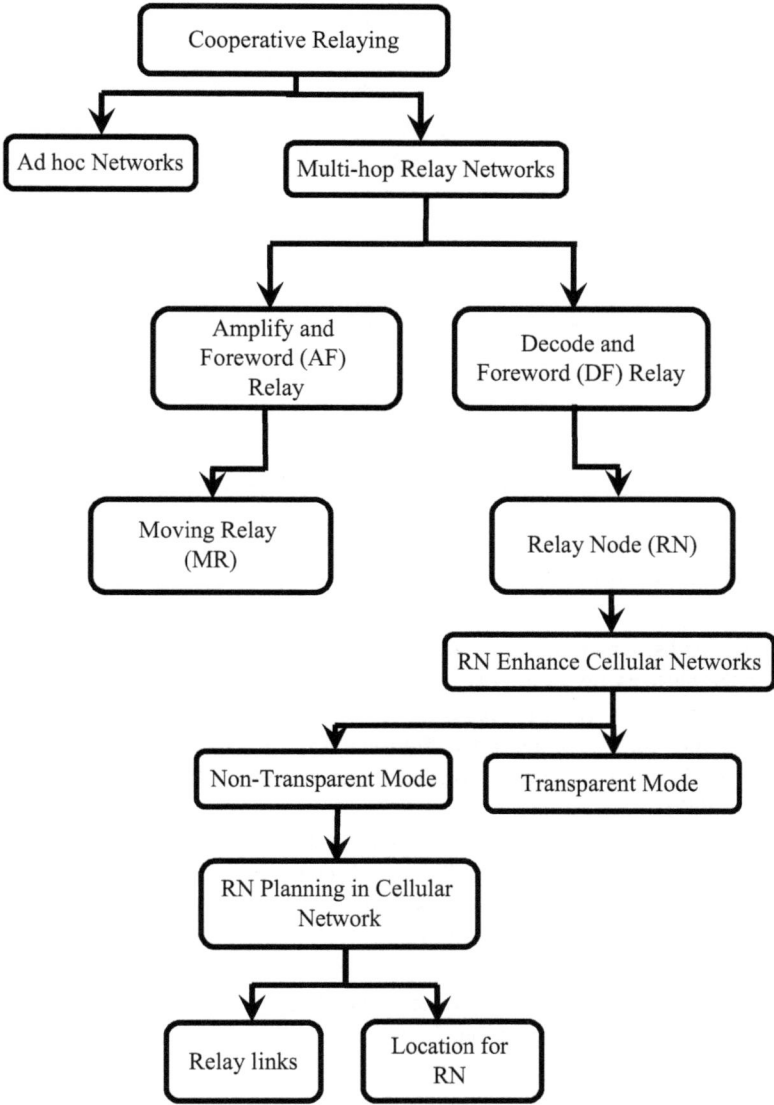

Fig. 2.3 Scope of this chapter

the User Equipment (UE) acts as a base and a receiver to route the information to its destination (Sharma & Jain, 2010). The following subsection explains in brief the Ad Hoc networks and multihop relay.

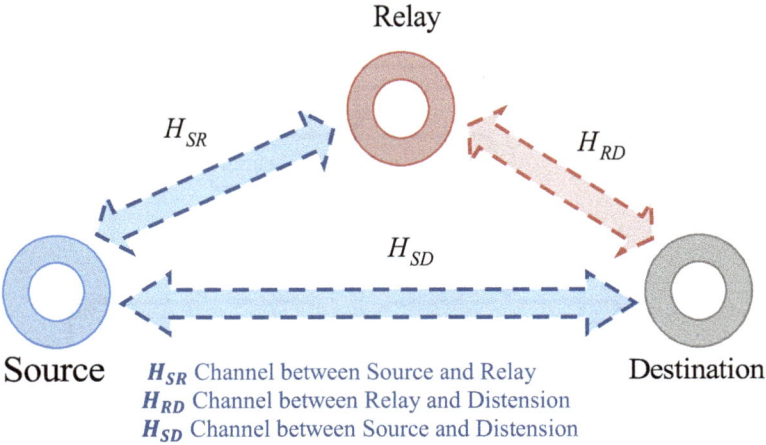

Fig. 2.4 Three terminals of cooperative relaying (Duong & Zepernick, 2009)

2.3.1 Ad hoc Networks

Ad hoc networks, which are also called mesh networks, are defined by the manner in which the network nodes are organized to provide pathways for data to be routed from the user to and from the desired destination. Ad hoc networks are basically those networks that have no Base Station (BS). In other words, every device (or node) in the network can act as a base and a receiver (Frodigh, Johansson, & LaRSSon, 2000).

A multihop ad hoc network consisting merely of mobile users is known as Mobile Ad Hoc Networks (MANETs),where every node can play the role of an intermediary node that relays messages from other nodes toward their destinations and cannot be accessed using a single-hop transmission (Sharma & Jain, 2010). MANETs include a group of mobile nodes that communicate without needing a fixed wireless infrastructure. Therefore, the communication between nodes is implemented by direct connection. Some nodes in the network are expected to help in the routing of packets and all hosts are allowed to move freely through the network. Successful routing protocols introduce methods providing packets to destination nodes given these dynamic topologies (Williams & Camp, 2002).

The Ad hoc network has many advantages. First, it does not require more fixed base transmitter. Second, the Ad hoc network is decentralized from the main network and is self-configured at the nodes and routers. Therefore, it has ability to self-heal by continuous reconfiguration (Williams & Camp, 2002). Finally, Ad hoc network is scalable to absorption of the addition nodes and is flexible to access the internet from different locations (Sharma & Jain, 2010).

The main disadvantage of Ad hoc network is that there should be nodes within the transmission range. If these nodes are not available, the whole network will fail (Frodigh et al., 2000). In addition, there is a lower data rate where the throughput is affected by system loading as well as for large networks, excessive latency (time delay) (Sharma & Jain, 2010).

2.3.2 Multihop Relay

Cellular systems conventionally employ single hops between UEs and the BS. The low throughput at the cell edge region has become a major concern to cellular network planners (Pabst et al., 2004). The effective solution to the problem of improving coverage and capacity is the use of small cells between source and destination (Li, Seet, & Chong, 2008).

Analog repeaters are used in cellular networks to help extend coverage to areas that the BS cannot cover. Drucker (1988) was one of the first to use the term cellular repeater and address the idea of using relays as repeaters to extend the coverage of an underlying cellular network.

In recent years, Multihop Relay-based networks have become an area of significant research interest in both industry and academia. Relaying information through multihops reduces the overall transmission power of the network without loss of reliability, thereby resulting in extended battery life (Seo, Mok, & Lee, 2007).

In order to meet with the rapid growth of wireless communications, which requires higher data rates and a more reliable transmission link while keeping a satisfactory Quality Of Service (QoS), researchers have focused on Multihop Relay scenarios (Duong & Zepernick, 2009). Multihop Relay is a low-cost solution, which provides a wide range of services in next-generation wireless networks. It can reduce the transmission distance and increase the number of users under channel conditions, for better link quality and higher throughput (Zheng, Lei, Wang, Lin, & Wang, 2011).

Yu et al. (2008) incorporated the multiple-hop scheme based on mathematical formulas to minimize the network cost and reduce the path losses between users and BS/RS. However, their proposed technique could not guarantee access quality for multiple-hop networks because authors relied on a random deployment of RSS within the cell. In addition, the location for RSS was chosen based on minimizing the network installation cost, regardless of the interference between stations and achieving the best enhancement in capacity (Prommak & Wechtaison, 2012).

Multihop Relay is one of enhancement keys which are introduced in LTE-A to improve the performance of 3GPP-LTE, in terms of capacity enhancement and coverage extension (Joshi & Karandikar, 2011).

2.3.3 Advantages of Multihop Relay

The concept of using Multihop Relay for wireless communication has been an active research topic for more than three decades. The main advantages of using Multihop Relay in cellular networks can be identified as follows:

1. *A reduction in total transmission power*: The total power consumption through the multihop relay is lower than direct transmission (Seo et al., 2007).

2. *An increase in network capacity*: Reducing transmission power leads to a reduction in the coverage radius of BS in Multihop Relay network compared with single hop. Therefore, the spectra can be reused more frequently as a result of the shorter reuse distances and the network capacity can be increased. This is similar to small cellular networks, where the coverage area is inversely related to spectral efficiency (Sharma & Jain, 2010).

3. *Higher throughput services*: Typically in cellular network, UEs near the BS are able to enjoy high throughput services, while those far away from the BS will have low throughput services due to power limitations. However, with Multihop Relay, the users far away from the BS can still access high throughput services whose data can be relayed via multihop networks (Boccardi, Yu, & Alexiou, 2009; De Moraes, Nisar, Gonzalez, & Seidel, 2012).

4. *Balancing traffic load*: Unbalanced traffic distributions complicate the management issue which allocates capacities in the network. This means that, some cells have enough available channels while other cells are heavily crowded, even though the traffic load has not reached the planned maximum capacity for the cellular network. Therefore, some users may be blocked due to the saturation in crowded cells in the network (Chiang, 2005). The Multihop Relay considered as a possible solution because it allows the traffic from congested cells to be transferred to other noncongested cells. Because of this, the probability of call blocking at the network point can be decreased due to load balancing between the cells (Chiang, 2005; Hyytiä & Virtamo, 2007).

5. *Mitigating capacity bottleneck*: The peer-to-peer communication process in multihop systems without involving the BS mitigates the potential capacity bottleneck that can rise due to the limited channels available to the BS in the single hop system (Lee, Han, Song, & Cho, 2006; Li et al., 2008).

6. *An increase in network coverage*: An improvement in capacity and an increase in coverage area in cellular networks are the major benefits of Multihop Relay networks due to a reduction in the path loss effect and an improvement in the Received Signal Strength (RSS) at UEs that are located in dead-spot areas of the cellular networks (Pabst et al., 2004). Dead spots may include the regions near the cell border, areas with deep fading (e.g., behind a building or in a tunnel), or areas where high interference prevents a clear reception of cellular signals (Sharma & Jain, 2010).

7. *An improvement in routing reliability*: In pure Ad hoc networks, the routing path is often vulnerable to node mobility and node failure, while in Multihop Relay the routing decisions can be controlled by intelligent BSs (Sharma et al., 2010).

2.3.4 Drawbacks of Multihop Relay

The advantages of Multihop Relay are also accompanied by several disadvantages that limit the use of the Multihop Relay technique. The main disadvantages are as follows:

1. *Increased network complexity*: Multihop Relay networks are small stations combined with main stations in the network. Therefore, network complexity is increased in terms of handover, routing, and resource allocation for peer-to-peer communications compared to single hop or the MANETs. In addition, the main BS may require a careful routing mechanism for a large number of UEs, with much larger than normal MANETs to exploit the benefits of relaying (Liu, Wan, & Jia, 2006).

2. *Increased interference*: The use of Multihop Relay will certainly generate extra intra- and intercell interference, which may potentially cause system performance to deteriorate.

3. *Delay*: At the multihop transmissions, packets are buffered at the relays before being forwarded to their destination. As a result, end-to-end delays are higher compared to single-hop transmissions, especially when congestion occurs due to high traffic loads (Wei & Gitlin, 2004).

2.4   Concept of Relay Node

Signal propagation through a wireless channel faces more constraints than a guided wire including greater additive noise, multipath, fading, cochannel interference, and adjacent channel interference (Rappaport, 1996). However, wireless transmission has become the appropriate platform to transfer information nowadays, due to the associated support, the freedom of the user from being physically connected and providing, portability and flexibility (Wyglinski, Nekovee, & Hou, 2010). The design of a reliable wireless system is difficult due to the random nature of the wireless channel and the diversity of environments in which they are likely to be deployed. Then next generation of wireless systems requires a higher voice quality as compared to current cellular mobile radio standards and has to provide higher bit rate data services with extension in coverage Dahlman et al. (2011a).

Recently, LTE-A technology has advanced and provides high capacity, low latency, and flexible bandwidth. Relaying is considered to be one of the main aspects of LTE-A which allows the system to meet the IMT-Advanced to improve 3GPP-LTE performance, in terms of coverage and throughput (Akyildiz, Gutierrez-Estevez, & Reyes, 2010; Khan, 2009).

Most wireless services in LTE provide enhancement in data service by adopting recently developed technologies such as multiple-input multiple-output (MIMO) antenna and by employing Orthogonal Frequency Division Multiplexing (OFDM) (Can, Yomo, & De Carvalho, 2007). However, in practice there are still issues such as coverage holes due to shadowing, and poor Signal to Interference plus Noise Ratio (SINR) for users that are far away from the BS (Yongchul & Sichitiu, 2011).

A solution for these problems is to add more base stations (BSs) with a small size coverage area; however, it is a very costly solution, especially when there are few users to be served (e.g., in rural areas). In addition, this solution increases interference (Abdallah Bou Saleh, Hämäläinen, Redana, & Raaf, 2012). An alternative to adding more BSs is deploying low-cost relay nodes (Rahman & Ernstrom, 2004) which provide a cost-effective way to overcome the problem (RNs are a simplified version of a full BS requiring lower cost than BS). Moreover, RNs do not require backhaul connections, thus reducing operating costs (Huang, Wang, Chang, & Su, 2010). Relay technology specification is considered by the LTE-A to meet the next generation of wireless communication technology, where it enables efficient use of communication resources, by allowing nodes to exchange information with each other to enhance the quality of services for current communication networks (Frederiksen, 2008).

An RN is connected wirelessly to the radio access network via a BS cell and receives, amplifies, and then retransmits the downlink and uplink signals to overcome areas of poor coverage within the cell. The RN is located either at the cell edge or in some other area where the coverage is poor. For multihop relay a downlink signal is sent from the BS to the RN and then to the UEs, while an uplink signal comes back from the UEs and is transmitted via the RN and back to the BS. The relay link between the BS and the RN caters for a growing number of users (Bulakci, 2012b).

The RN helps to resist the propagation loss through a division of the path loss between source and destination into two parts, thus the path loss in the sum of two parts is less than the path loss in the whole path. This feature of the relay technique reduces the impact of path loss and can be referred to as path loss gain. Theoretically, SINR is inversely proportional to the signal propagation distance SINR $\propto 1/d^{\alpha}$ where d is the distance between the BS and UE and α is the path loss exponent, which typically ranges between 2 and 6 based on the type of the propagation environment (Sesia et al., 2011). When the RN is midway between the BS and UE and the transmitted power (p_t) is divided equally between the BS and the RN, the gain of received power (p_r) compared to the traditional point-to-point system, as shown in Fig. 2.5, is determined by the following equations (Chen, 2012).

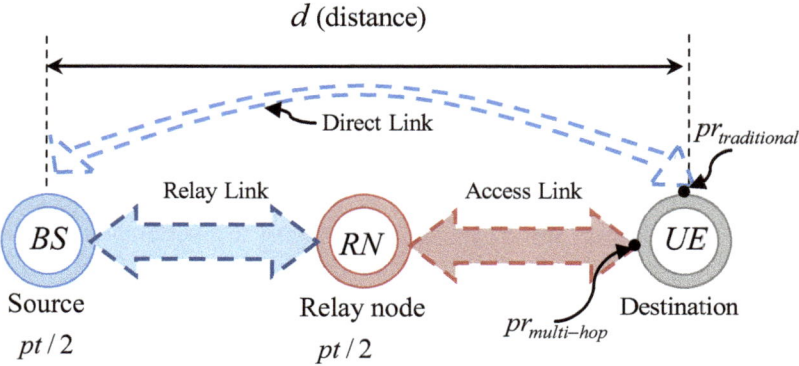

Fig. 2.5 Typical scenario of multihop relay

$$p_r = p_t \cdot d^{-\alpha} \tag{2.1}$$

$$\frac{p_{r\ \text{multi-hop}}}{p_{r\ \text{traditional}}} = G_p = \frac{\frac{1/2(p_t)}{(d/2)^\alpha} + \frac{1/2(p_t)}{(d/2)^\alpha}}{1(p_t)/d^\alpha} = 2^\alpha \tag{2.2}$$

$$p_{r\text{multi-hop}} = 2^\alpha p_{r\ \text{traditional}} \tag{2.3}$$

Relay offers several advantages and disadvantages for mobile communications (Sharma & Jain, 2010; Wyglinski et al., 2010), which are listed as follows:

2.4.1   Advantages

There are many advantages associated with relay node deployment within LTE cellular network and stimulated researchers to use relaying technique. The main advantages are as follows:

1. *Coverage Extension*: Coverage of the cell is affected by signal transmission loss, especially a user at the cell edge (Guo, Wang, & Chu, 2013). However, a relay node can effectively expand the network coverage by signal amplification.
2. *Quality of Service (QoS):* The Relay system is effective in resolving the effects of channel fading by cooperative diversity and effectively enhancing transmission robustness by guaranteeing the transmission between the BS and users (Huang et al., 2010).
3. *Capacity Enhancements*: For cellular networks which have a wide coverage area, high performance gains can be achieved using relay by dividing the path loss into two or more hops. These gains provide a higher capacity and transmission rate (Chen, 2012; Wyglinski et al., 2010).

2.4.2   Disadvantages

The advantages of relay in cellular networks are that they rely on location for the RN in the cell (Khakurel, Mehta, & Karandikar, 2012; Meko, 2012).

1. *Increased Interference:* Using relays introduces extra intra- and intercell interference, which potentially causes the system performance to deteriorate (Zhao, Fang, Huang, & Fang, 2014).
2. *Network Complexity*: Relays increase network complexity in terms of handover, overlapping, and resource allocation for peer-to-peer communications, as compared to single hop.

A new type of relay called Moving Relay may be installed in vehicle (Sui, Papadogiannis, & Svensson, 2012a). MR improves throughput for passengers in urban areas where there is a high shadowing effect due to buildings, as well as in rural areas, where there is a weak signal from BSs, especially at the cell boundaries (Bulakci, 2012a; Sui et al., 2012a). The disadvantage of MR is that transmission continues irrespective of the high received signal strength from BS at users (Bulakci, 2012a). This work proposes a method to minimize transmission power from the MR.

2.5   Relays Classification

Relays can be classified according to the signal processing techniques employed and the functionality of relay in the Amplify-and-Forward (AF) relay and the Decode-and-Forward (DF) relay.

2.5.1   Amplify-and-Forward (AF)

The basic function of the AF relay is the amplification of the received signal from the source and a retransmission of the same signal to the destination without processing. The AF relay not only amplifies the desired signal but also amplifies interference and noise which deteriorates the overall SINR level as well as limits the system throughput.

The AF relay introduces low delay due to filtering, processing, and feeder links used in its application which deteriorate the signal coming to the user terminal. However, there are weak points with this type such as an increase in the noise and interference level in the system, and an accumulation of erroneous data over multiple links (Bletsas, Shin, & Win, 2007; Yang, Hu, Xu, & Mao, 2009) as shown in Fig. 2.6.

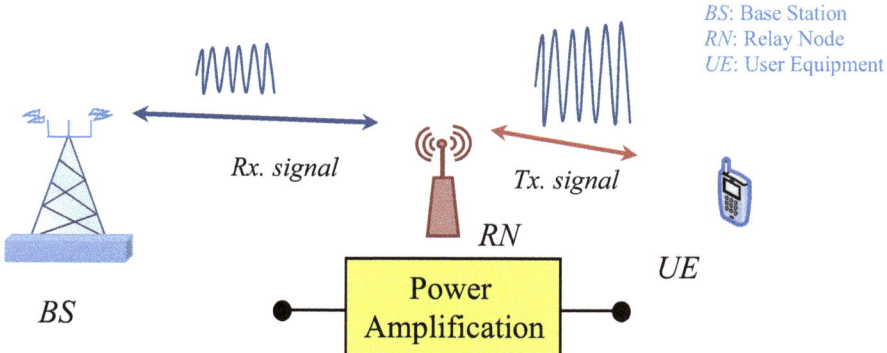

Fig. 2.6 Amplify and foreword relay

Researchers have discussed increasing the coverage in cellular networks through the use of AF relay. Jeon et al. (2002) and Rahman and Ernstrom (2004) studied the effects of installing AF relay in hotspots in Code Division Multiple Access-2000 (CDMA2000) for 3G cellular networks. The authors based the relation of the QoS system with both capacity and coverage analysis by using AF relay as repeater between the BS and users. The simulation results proved that there was improvement in downlink coverage, especially for hotspots which were located halfway between the BS and the cell border. However, the authors did not discuss increasing of the coverage for the whole cell as well as the interference and power allocation for stations has been neglected in the study.

The protocol of AF relay is lower complexity than all cooperative strategies; more studies are proposed to maximize the achievable rate for AF relay (Fei, Qinghua, Tao, & Guangxin, 2007). Laneman Tse, and Wornell (2004) proposed AF relay of low complexity protocols for cooperative diversity. In their proposed scheme, a simple AF relay for the half-duplex mode was analyzed where the sender and relay had equal power constraints. The authors improved the performance descriptions for AF relay in terms of outage statuses and associated outage probabilities. The study was based on measured robustness of the transmissions against the fading channel with a high considered SNR for the system. The performance characterizations for the scheme showed that large power savings result from the use of these protocols.

Rizinski and Kafedziski (2011) introduced a scheme to achieve maximum rates of the AF relay strategy for the Gaussian relay channel. Both full-duplex (FD) and half-duplex (HD) modes are proposed with the proposed channel. The numerical results showed that the AF strategy achieves highest rates when the relay is located midway between the transmitter and the receiver. However, these studies are based on assumption of perfect transmission at the transmitter and relay so that the channel conditions of direct and relay links are assumed equal. As a result, the results were ideal. In addition, interference and power allocation were not considered in the studies.

Fei et al. (2007) and Gurrala and Das (2012) studied the impact of relay location in terms of Sample Error Rate (SER) performance. The effect of relay placement was studied according to the SER performance analysis through a simple line topology. Transmission powers allocated for source and relay are assumed equal. Simulation results indicated that the maximum rate is achieved when AF relay is located halfway between source and destination. However, the mathematical analysis of the study was based on a single relay in the cell as well as interference between the stations is ignored, thus giving unreliable results.

Wirth, Thiele, Haustein, Braz, and Stefanik (2010) introduced an approach to improve capacity and RSS within an indoor office scenario with different locations to install the optical Distributed Antenna System (DAS) for AF relay. Simulation results showed there was an improvement in RSS and throughput inside the building after using relay and when the antennas were installed in the four corners of the room. The drawback in the study is that increasing the self-interference among the antennas in addition to using DAS for AF relay leads to increased cost and complexity of the network by combining the fiber systems with RF devices.

2.5.2   Decode-and-Forward (DF)

DF relay fully decodes and reencodes the received signal before the retransmitting to the destination (Duong & Zepernick, 2009). This scheme provides advantages for radio resource management and is employed in interference-limited environments. However, these proceedings of the signal require some time at the RN which leads to delay and an increase in system complexity, as shown in Fig. 2.7.

Host-Madsen and Zhang (2005) introduced the approach of three-node wireless DF relay channels in a Rayleigh fading environment for the upper bounds and lower bounds on the outage capacity. The scheme showed the comparison between the direct transmission and traditional multihop protocols for a variety of wireless relay channel models. The results showed that optimal relay channel signaling can significantly outperform multihop protocols, and power allocation can yield a significant gain in wireless relay channels. Although the study addressed important issues for the optimum DF relay channel, the work is a theoretical study for DF relay channel performance apart from the interference and resources allocation if used with cellular networks. As a result, it may not provide high-precision results (Ng & Yu, 2007).

DF relays are currently being specified in 3GPP-LTE work to provide the LTE-A network requirement in order to meet the growing demand for coverage extension and capacity enhancement (Iwamura, Takahashi, & Nagata, 2010).

Abdallah Bou Saleh et al. (2012) investigated the performance of DF relay for in-band and out-band within the LTE-Advanced for different propagation scenarios in terms of both coverage extension and capacity enhancement. In order to study the effect of the relaying overhead on system performance, the study compared the in-band and out-band operation mode for DF relay. Simulation results showed that in-band relay deployment offered low to very high gains compared to out-band based on the differences of the environment. However, the spectrum sharing and power allocations have not been taken into account in the study. On the other hand, should allocate the frequency band between relays and BS when using in-band operation mode to reduce the interference between the nodes.

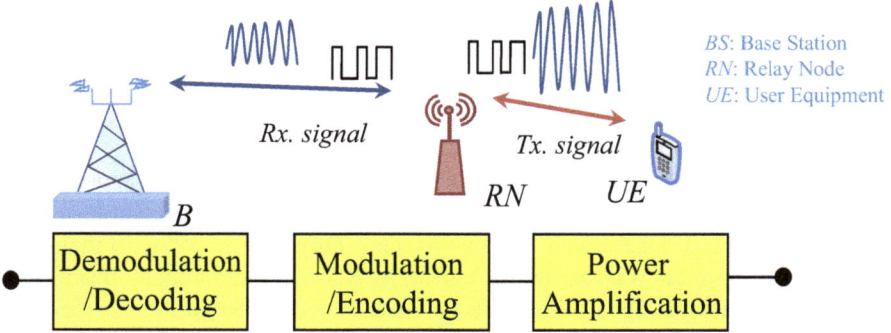

Fig. 2.7 Decode-and-forward relay

Zhao et al. (2007) analyzed the coverage extension of LTE-A cellular networks using DF relays. They showed the relationship between the number of relays and the coverage range extension based on the coverage angle and coverage range for relay with BS. The coverage is determined by moving one UE around the size of BS and then receiving the data from the nearest relay. The authors assumed the transmitted power allocation for different numbers of antenna configurated in BS and RNs was equal. The simulation results indicated the coverage range extension for MIMO relaying systems provided a huge improvement as compared to direct transmission. The main drawback in their work is that the orderly distribution of relays, regardless of the optimal placement of the relay that provides the best coverage and ignores the fact that allocating the transmission power for each relay could exacerbate interference among the stations.

Rizinski and Kafedziski (2011) analyzed achievable rates and capacity bounds for the Gaussian relay channel by using DF relay within decoding and without decoding to channel. The authors investigated the effect of various channel states of the relay links (sender to relay, relay-receiver) and direct link (transmitter–receiver) on the achievable rates. The numerical results indicated the DF strategy with decoding was preferable when the relay was close to the transmitter, while the DF strategy without decoding was preferable when the relay was close to the receiver. In addition, using multihop strategies in wireless provides high bit rates compared with direct transmission. Although the study presented the analysis of DF relay channels with different strategies, the supposed transmission power for both relay and BS being equal could give unrealistic results.

The following section explains other relay categories which are considered by LTE-A networks to enhance 3GPP-LTE networks. The first is known as Relay Node (RN), while the second is called Moving Relay (MR).

2.6 Relay Node (RN)

The Relay Node (RN) and sometimes called Relay Station (RS) is a fixed small cell usually deployed in dead and crowded regions (Yongchul & Sichitiu, 2011). This means that the RNs are deployed to the poor coverage area of the cell to enhance the capacity of users in these areas (Salem et al., 2010). These areas might not necessarily be on the cell edge but in highly shadow-faded areas like behind large obstacles or inside or behind buildings. RN can also be deployed to enhance the existing capacity of certain areas, such as a busy street in a city center or an indoor office. These areas are called hot spots as shown in Fig. 2.8.

RNs are connected to the power supply unit, thus allowing them to have high access to processing capabilities and possibly higher transmission powers than regular UE, which can ease the process in finding the capacity enhancement in relaying (Meko, 2012). The location of the best spots for the RN can be found either via measurements, simulations, or demand (Khakurel et al., 2012). The RN can have fixed antenna higher from the ground level toward the BS to enhance the radio

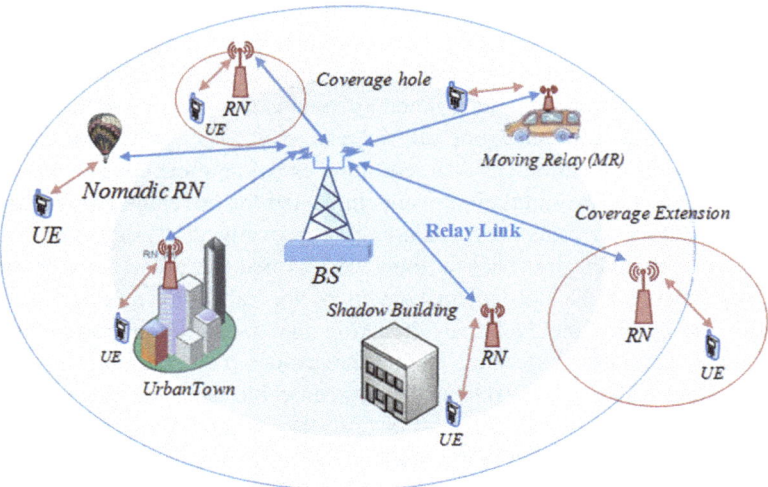

Fig. 2.8 Scenarios of relay nodes

links and increase throughput. MIMO capabilities may be also easier to implement in RNs when the antennas and radio are fixed in one place (Ding, Krikidis, Thompson, & Leung, 2011; Xu, Dong, & Lu, 2011).

In recent years, the interest in RN has increased with the rapid growth of technologies which require high data rate irrespective of the user's location such as LTE-A cellular networks in order to meet with 4G requirements.

Vidal, Marina, and Host-Madsen (2008) analyzed the DL spectral efficiency of multicellular networks so that each cell consists of BS and a certain number of deployed RS. HD mode is proposed in the study where the BS transmits the signals either direct or via RS in two hops. Based on the classical Shannon capacity expressions, the authors discussed the impact of the position of the RS which enhances DL spectral efficiency. Simulation results showed for a fixed cell radius the spectral efficiency doubled compared to direct transmissions using a single antenna at the BS, RS, and UE. The study demonstrated the enhancement in the capacity for the cell by deploying RS based on Line Of Sight (LOS) transmission. However, the authors assumed a uniform distribution of RS in the cell and did not discuss the interference and resource allocation issue, which are the main challenges facing relay deployment for cellular networks

Martins, Rodrigues, and Vieira (2012) showed a way to determine the location for RS in the coverage size of BS and also quantified the associated performance gain using different cluster size configuration. In the work, the relays were deployed within selected area. The system performance was tested with a different number of relays according to path loss variations. The simulation results indicated that there was an enhancement in coverage when the number of RS increased. Although, the increasing number of RSS in cell increases the coverage and capacity, it also exacerbates the interference between stations and adds extra cost to the network.

Moreover, determining the position for relays in the work relied on the selection of the best position for RS among predefined positions. As a result, the work did not provide the selection of optimal RN locations within overall cell dimensions. Survey for relay node in details is reviewed in Sects. 2.7 and 2.8. The following section shows relay node enhancement in cellular networks.

2.7   RN Enhance Cellular Network

Cellular systems have evolved dramatically in recent decades with the advent of new and efficient radio access technologies, and the implementation of advanced techniques to support the increments in capacity and coverage in cellular networks (Akyildiz et al., 2010). Although great efforts have been made in the research community, the traditional single hop cellular architecture fails to provide high and fair QoS across the cell area. Because the effect of propagation loss does not guarantee a fair distribution of the signal strength across the cell, it means that there are large areas which have a low signal level, especially at cell boundaries (Joshi & Karandikar, 2011).

One of the solutions to this problem is increase the number of base stations, so that station is able to cover every small area. However, increasing the number of BSs requires a high deployment cost and causes higher intercell interference. Hence, a cost-effective solution is needed to cover the required area while providing the desired high SINR to users meeting with the demand of future cellular networks (Meko, 2012).

Using multiple hops from source to destination reduces the communications distance and signal loss in each hop, and thereby gives the ability to increase the overall multihop transmission capacity compared to long-distance and single-hop communication links (Pabst et al., 2004). In addition, the use of relay nodes improves signal quality by replacing NLOS links by multiple hops with LOS propagation conditions (Muñoz, Coll-Perales, & Gozalvez, 2010). Recently, there has been increasing interest in integrating Multihop Relay functionalities into cellular wireless networks (Lin & Ho, 2007).

Dinnis and Thompson (2007) investigated the enhancement in the DL data rate and coverage in randomly positioned UEs of a cellular network using RNs. The SNR was evaluated and a channel in the model was based on both path loss and lognormal shadowing for relays deployed in the cell. The numerical results showed improvements in coverage and capacity with relay deployment so that the amount of this improvement depended on the relay transmit power and relay density in cellular networks. However, the results were based on the assumption that there was no interference between the different nodes.

Viswanathan and Mukherjee (2005) proposed a centralized DL scheduling scheme in a cellular network with a small number of relays. The scheme guaranteed the stability of the user queues for the largest set of arrival rates based on the potential gains which were achievable in a relatively idealized setting (including a simple model for the signaling overhead). Simulation results indicated larger gains achieved in terms of throughput where the relays were placed in hot spot locations. The drawback of the scheme is the consideration that in the ideal environment there is no interference between the nodes based on complex algorithms. Besides, deploying a small number of relays irrespective of the optimal location could provide unrealistic results.

Madan et al. (2010) describe schemes for designing the heterogeneous cellular networks such as pico/femto BSs and relays in a macrocellular network in terms of cell splitting, range expansion, and resource allocations. The authors discussed the fair distribution of resources based on the concept of a concave utility function of average rate. In addition, they designed simple fast distributed mechanisms and algorithms for dynamic interference management for every subframe. Numerical results showed that there were large gains over currently used methods for cellular networks. Although this study introduced important concepts for heterogeneous cell deployment within cellular networks, the optimal locations of each small cell and appropriate number required in the cell to ensure best performance for the network was not addressed in the study.

Krishnan et al. (2012) investigated bandwidth allocation in LTE-A cellular networks employing AF relays, where they studied the jointly optimized bandwidth and power usage under constraints such as the required rate, bandwidth, and transmission power. They also showed specific results with many users in order to allocate bandwidth and power. Numerical results indicated that transmission power savings were at least 3 dB, which could be achieved by optimizing both bandwidth and power. However, this scheme did not address multiple cell and intercell interference, which is considered to be an important issue in relay deployment scenarios to achieve realistic results.

In general, it can be stated that most of the earlier studies discussed improving at cellular network performance by deploying RNs. Nevertheless, these approaches proposed the random distribution of relays over cells with channel selection according to path loss and SNR and did not discuss the effective cost and power allocations for the relays. The following section explains relay mode operation which is considered as a standard of LTE-A specifications.

2.8 RN Mode Operation in LTE-A

Two of the operational modes considered in LTE-A standard are the transparent and nontransparent mode (De Moraes et al., 2012). These modes are employed based on the purpose of usage, whether for throughput enhancement at the cell edge for users or for coverage extension. Therefore, the modes of operation at Multihop Relay for LTE-A cellular network may be classified into transparent and nontransparent.

2.8.1   Transparent Mode

In this operational mode, a UE connected to a RN is located within the coverage of the BS and the control signaling from the BS can be accessed directly to the UEs, while the data traffic is relayed via the relay node. Therefore, the control signaling and the data traffic are separated (Genc, Murphy, & Murphy, 2008; Kwon, Chang, & Copeland, 2008) as illustrated in Fig. 2.9.

BS coordinates and allocates the radio resources to UEs and RNs within the cell coverage, by guiding and distributing the control information and access requests. Transparent mode is dedicated to throughput enhancement inside buildings and congested areas where the UEs are located within the coverage of BS (Zeng & Zhu, 2008).

Recently, the rapid growth of wireless services has required high throughput to users: Thus, the multihop with transparent relay type provides a high data rate inside crowded buildings. In this regard, Genc et al. (2008) investigated system throughput enhancement which can be provided by the deployment of transparent relays in a Worldwide Interoperability for Microwave Access (WiMAX) network. Later Genc, Murphy, and Murphy (2009) studied the network capacity of IEEE 802.16j systems operating in transparent mode with varying numbers of relays and incorporated transmission power. Satish Kumar and Nagarajan (2012) proposed a new adaptive model to improve throughput and to select appropriate relay node according to the optimal relay selection procedure. The results indicated improvement in throughput by placing a transparent mode relay. However, resource allocation such as power and frequency as well as the issue of interference was not considered, meaning that the results were unreliable.

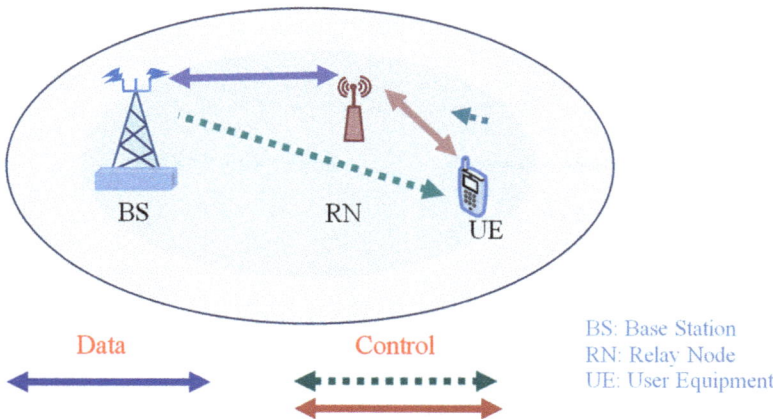

Fig. 2.9 Transparent mode

2.8.2 Nontransparent Mode

In nontransparent mode all data and control signaling transmissions between the
BS and UE are relayed via RN. Therefore, the RN operates in both centralized
and distributed scheduling and has ability to extend the coverage as illustrated
in Fig. 2.10. Table 2.2 presents the comparison between transparent and
nontransparent mode.

The comparison of network coverage and throughput (cell throughput) for
multihop relay networks with single-hop system (as a baseline), transparent and
nontransparent, were studied by (Zeng & Zhu, 2008). The authors paid special
attention to simulate the multipath fading channel and interference among BSs or
RNs in DL transmission. However, the study made an important comparison
between transparent and nontransparent relay modes in terms of both scheduling
distribution and the adaptive modulation scheme based on the channel variations.

Kyungmi et al. (2009) investigated bandwidth efficiency and the associated
service outage performance for different nontransparent relay mode scenarios
using system level simulation for a cellular Orthogonal Frequency Division

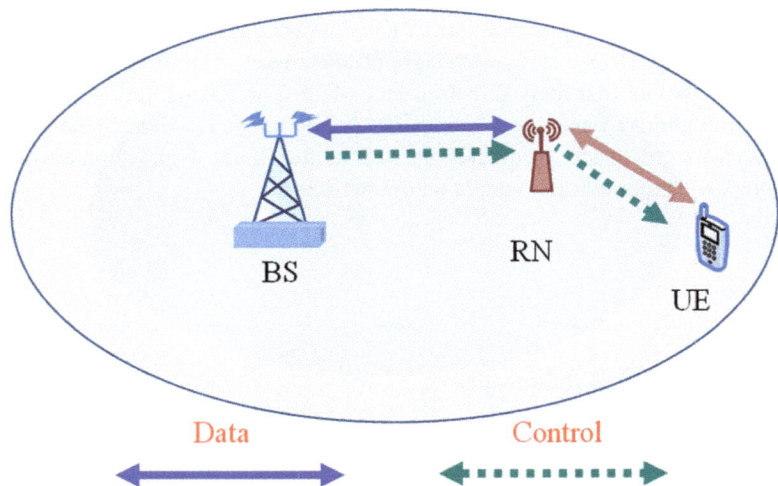

Fig. 2.10 Nontransparent mode

Table 2.2 Comparison of transparent and nontransparent mode (Akyildiz et al., 2010)

	Transparent	Nontransparent
Coverage extension	No	Yes
No. of hops	2	2 or more
RN cost	Low	High
Scheduling	Centralized scheduling only	Centralized/distributed scheduling

Multiple Access-Time Division Duplexing (OFDMA-TDD) system. The authors showed that multihop relays increase the coverage and system capacity by using a simple AF relay. In addition, the simulation results demonstrated improvements of data rate per user at the cell edges. Although the results provided enhancement in coverage, the number of relays and their placements was not discussed, whereby system performance remains unfair in terms of coverage distribution within the overall cell.

2.9 RN Planning in Cellular Network

Conventional cell planning includes positioning the BSs, determining the power level of each BS, and aggregating radio frequency among BSs antennas. The optimization objective is to minimize the total deployment cost or maximize the number of demand nodes with a given deployment budget (Amaldi, Capone, & Malucelli, 2003).

Relaying technology introduced in LTE-A provides high link quality and achieves high data rates for the users in 3GPP-LTE cellular networks. In addition, RN deployment cost is lower than BS where RNs are smaller in size, there have lower power outputs, and also there is the possibility to install it on preexisting lamp posts (De Moraes et al., 2012; Doppler et al., 2007; Joshi & Karandikar, 2011; Khakurel et al., 2012).

Guo et al. (2010) investigated relay deployment to assist disaster area networks for first responders as user nodes. They studied the relay management problem of finding a minimum number of RNs and their dynamic locations to cover all the user nodes within the disaster area. The authors proposed an algorithm to control the amount of relay deployment with a transmission range of user nodes in a disaster area. The simulation results indicated that the number of relay increases as the transmission range of user nodes decreases. However, the approach provides for RN deployment within specific areas known as disaster areas and did not address the frequency reuse issue and power allocation for each RNs. Moreover, the optimal location of the relay within the BS was not discussed. Therefore, the study is appropriated for special cases when increased traffic is suddenly required in special area.

Abdallah Bou Abdallah Bou Saleh et al. (2012) and Bulakci, Redana, Raaf, and Hamalainen (2011) studied relay site planning by improving the quality of the relay link. The impact of site planning in terms of two proposed techniques such as cell selection and location selection on the system and end-to-end rate has been investigated in order to enhance overall system performance. The study was based on the channel quality of relay links to deploy RNs within the cell size. Results demonstrated significant improvements in bit rate which justify the need for relay site planning in relay enhanced networks. The study highlighted limitations of the relay link and showed the potential for significant extra gain if these limitations are relaxed. However, the results are incomplete for RN planning with cellular networks because the optimal location for the RN and the number of RNs was not discussed in order to avoid interference and provide more realistic and authentic results.

The relay deployment concept and overview of many parameters such as propagation and physical layer challenges, interference effects on the relay planning within cellular network were investigated in detail by Damnjanovic et al. (2012) and Pabst et al. (2004)

In order to ensure the optimization for relay deployment in the cell, it is necessary to specify the number of relays and optimal locations for relays which provide higher and fair QoS that mitigate interference problems (Genc, 2010). The following subsections present some of these aspects.

2.9.1   RN Location

With the increasing number of cellular subscribers, cellular systems are facing difficulties in providing satisfactory SINR level to users especially at the cell boundaries. One of many solutions to support the ever increasing number of subscribers per cell is to decrease the cell radius. However, this requires more base stations per area, thus increasing cost and causing higher intercell interference between stations (Joshi & Karandikar, 2011).

An encouraging solution is employed in LTE-A cellular systems by deploying low-cost RNs in each cell. The deployment of RNs has many advantages: capacity enhancement, coverage extension, and cost reduction (Khakurel et al., 2012; Meko, 2012). However, these benefits are based on the location of RNs in the cell. Therefore, it is very important to study the impact of optimal location for the RN in terms of link reliability and system capacity (Huang et al., 2010). When RN is deployed far away from the BS, the users at the cell boundary can receive a stronger signal to improve communication reliability, but the longer hop distance between the BS and RN leads to degradation in relay link capacity and increases the interference level among cells. On the other hand, placing the RN close to the BS causes more outage for users at the cell boundary and does not achieve the required objectives. Determining the appropriate location of relay provides the trade-off between communication reliability and system capacity (Khakurel et al., 2012; Yongchul & Sichitiu, 2011).

To date, only a few researchers have addressed the issue of optimal RN location in cellular network (Joshi & Karandikar, 2011; Lin & Ho, 2007). The cooperative relaying system with distributed Alamouti code is investigated by Lin et al. (2009), AF relay, DF relay, or hybrid AF/DF relay protocols were used in this scheme, and the performance of the system was based on optimal relay locations for different protocols and modulation schemes. Herein the relay location varied with selected modulation schemes and the effect of channel environment. Simulation results showed that the best performance for AF protocol was achieved when the relay location was 0.55 % of the distance between the source and destination with a Binary Phase-Shift Keying (BPSK), Quadrature-Phase-Shift Keying (QPSK), and

16-Quadrature-Amplitude Modulation (16-QAM) modulation scheme. However, the relay location varied from 0.53 to 0.6 % of the distance between the source and the destination with a selected modulation scheme by DF and AF/DF relay protocols. Although this work addressed the relay location for different modulation schemes and showed that the relay location depends on selected modulation schemes, the work did not study power distribution, interference, or the number of relays that provided integrated and realistic results.

B. Lin and Ho (2007) introduced network dimensioning and location planning (DLP) in a multihop wireless network model, by identifying a model of dimensioning and location planning of RSS and integrating the multilevel of cooperative relaying in the network design. In the model, the authors considered RS placement, relay allocation, and relay sequence design together in a unified framework. Two algorithms, the single and multihop system, were proposed in the model. Simulation results demonstrated significant cost reduction and achievable rate improvement due to the multilevel cooperative relaying rather than the single hop network. In the study, the authors discussed the cost effectiveness of multihop networks which employed relays and compared them with traditional planning. It is worth mentioning here that the works were based on complex algorithms to select the relay location as well as the optimal number of relays per cell related with interference mitigation which was not addressed in the study.

Esseling, Walke, and Pabst (2004)) and Pabst et al. (2005) showed that the analysis related to the traffic performance for a wireless broadband system was based on a fixed relay acting as wireless bridges. The authors discussed three typical resource allocation schemes for relaying networks including relaying in time domain, frequency domain, and hybrid time/frequency domain schemes. The analysis focused on the end-to-end performance in terms of throughput and delay. In addition, the scheme discussed both densely populated urban areas and wide-area environments to provide a coverage area within the 3GPP cellular network. Simulation results showed that an enhancement in capacity and broadband radio coverage in cellular wireless broadband systems through relay. The authors established that a suitable concept of fixed relay contributed substantially to providing high capacity and coverage through a relay which was deployed in next-generation cellular wireless broadband systems. However, the relay placement problem that provides maximum gains at the cell edges was not debated. In addition, the number of relays per cell which is related to the total cost and interference was not studied in these works.

Laneman et al. (2004) developed and analyzed low-complexity cooperative diversity protocols to combat fading and multipath propagation in wireless networks. They employed strategies of cooperating radios including relay such as AF relay and DF relay schemes and selected relaying schemes based on SNR. The relay selection scheme adapts based on channel measurements and outage probabilities which measure the robustness of the transmissions to fading. It should be noted that the study discussed important issues related to relay strategies in terms of downlink signal analysis and spectral efficiency based on complex algorithms.

Sadek, Han, and Liu (2010) developed the hybrid version of the incremental and selection relaying protocols proposed by Laneman et al. (2004). They addressed the relay-assignment problem for coverage extension based on the knowledge of the channel statistics governed by the user distribution in the cell. The authors analyzed the performance of two schemes: a distributed nearest neighbor relay assignment in which users can act as relays and an infrastructure-based relay-assignment protocol in which relays are deployed in the network to help the users to forward their data. The outage probabilities of the two schemes were derived. The optimal relay position was characterized to minimize the outage probability for the user. Simulation results showed significant gains when applying the proposed algorithm over direct transmission in terms of coverage area and spectral efficiency. The weakness in the work is that the number of relays determined previously with six RNs deployed in the cell and the assumption that the transmission power is equal for both BS and relay, which leads to overlapping among stations; especially interference has been neglected.

Dong, Zhang, Song, Teng, and Man (2009) and Wang et al. (2008) investigated the impact of relay location on the system capacity. Received data from the relay is determined according to both Received Signal Strength (RSS) and achieved throughput. The results showed the achievable throughput at UE via RS can be higher than that in the direct transmission at some UE locations. Based on the assumptions of Wang et al. (2010) proposed the optimal location of relay aims to maximize system capacity. Two relay selection rules: signal strength-oriented and throughput-oriented selections were adopted to determine whether multihop transmission would be used. Simulation results showed there was an improvement in throughput when increasing the number of relays based on the assumption that relay placement is fixed. However, the authors proposed K^{th}. RS regularly deployed around the BS and did not discuss the relation between number of relays per cell and the power allocated for relay which exacerbates interference and performance deterioration.

Finally, Meko (2012) studied the optimal relay placement for DF relay in cellular networks. The study based on evaluation of BS-UE, BS-RN, and RN-UE links through both signal strength and SNR as shown in Fig. 2.11.

The author considered a cellular system with six RNs that were placed symmetrically around the BS and evaluated the optimal relay placement based on the reference level of outage probability given by Joshi and Karandikar (2011) and Khakurel et al. (2012). Simulation results showed optimal relay placement schemes significantly reduced the outage of the cellular system and provided better QoS for users at cell boundaries. Although the technique showed substantial gains in terms of capacity, it should be remembered that the number of relays is specific and the relays are distributed regularly around BS irrespective of power allocated for each relay which does not provide the desired objectives of achieving the fair distribution of capacity at the cell edge. The work did not discuss intracell interference among the stations, which is an important issue in the relay deployment.

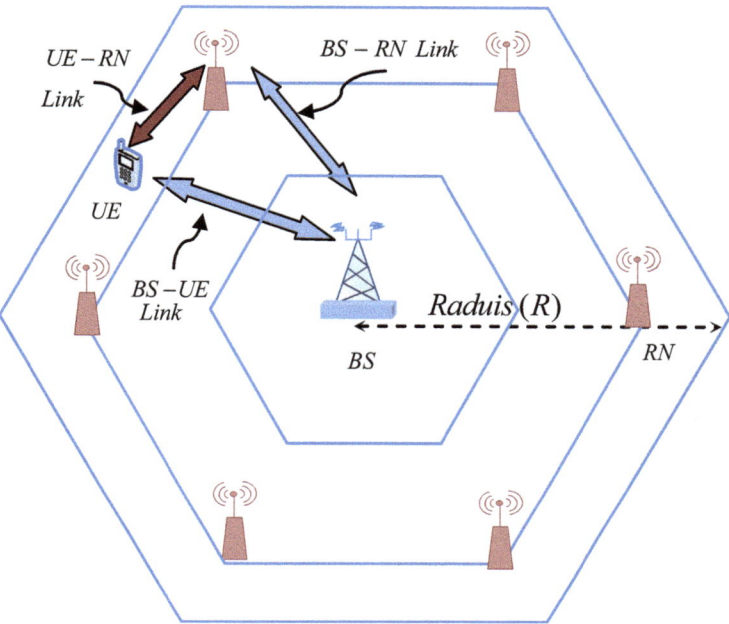

Fig. 2.11 Layout of RN enhanced cellular system and illustrating the coverage of UE-BS, RN-BS, and UE-RN links (Meko, 2012)

2.9.2   *Relay Links*

Relaying improves the coverage of cellular networks in LTE-A by providing high data rate coverage in the cell edge region, reducing the deployment costs of cellular networks, and improving the effectiveness of cell throughput and capacity. In order to minimize the cost of installation, the fixed microwave link (i.e., LOS microwave systems), the relay node was designed as wireless and connected with both BS via relay link and UEs via access link. Relay link provides flexible, cost efficient, and less time to deploy RNs in the cell (Han & Wang, 2010; Sadek et al., 2010). In LTE-A, the relay link is designed to operate with an in-band or out-band transmission mode. In the in-band mode the relay link operates in the same frequency band as the access and direct links (De Moraes et al., 2012) as shown in Fig. 2.12.

However, the in-band mode causes interference between the access and relay links when the relay link shares the same spectrum of access link. In order to avoid the interference of two links, additional mechanisms are required such as separation in the time domain between the relay and access links, so that the two links cannot operate simultaneously (Dahlman et al., 2011a). In out-band relaying mode, the relay link operates on a different frequency band than the rest of the cellular UEs. As a result, the interference between the relay and access links can be avoided and the necessary isolation be obtained in the frequency domain (Dahlman et al.,

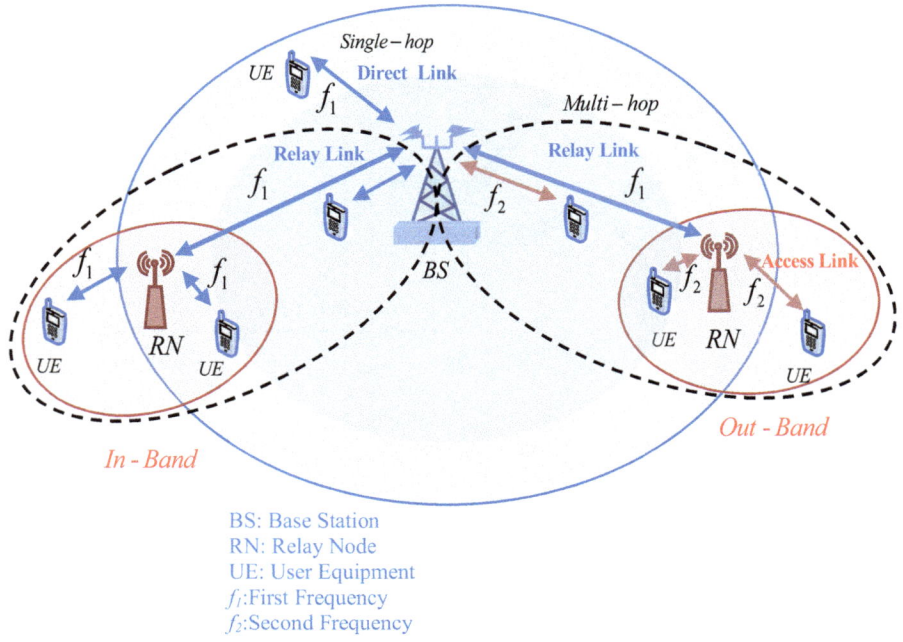

Fig. 2.12 In-band and out-band relay mode for LTE-A multihop network

2011b). However, this type requires separate RF filter and antennas which are expensive compared to in-band mode as shown in Fig. 2.12.

The long distance between the relay location and BS improves the coverage at cell boundaries but degrades the relay link capacity, thereby increasing the probability of outage. On the other hand, the approximating of the relay location does not achieve the desired goals for RN deployment. Therefore, any improvements in link quality ensure the capacity and required throughput for the growing number of users at the cell boundaries. So far there has been little research focusing on improving relay link throughput or introducing effective solutions to solve the problem (Kitayama, Hasegawa, Taniguchi, & Nakano, 2013).

Coletti, Mogensen, & Irmer, 2011a, 2011b introduced a study, which evaluated and compared the potential of LTE relay and micro deployment in a realistic metropolitan scenario. The authors proposed an algorithm of both modes (In-Band/Out-band) in relaying which combines both network coverage and a realistic spatial of user density information. The results indicated that for the downlink, the network improved in terms of coverage and reduced the user outage service values for users at in-band operation mode, whereas at out-band the capacity improved with a lower frequency carrier at the BS. However, these results were based on postulating a high quality at the relay link and the authors did not mention how to improve relay link quality which enables blocked users to access the network, especially at the cell edge.

A comparison of two operations, in-band and out-band modes in LTE-A networks was introduced by Gora and Redana (2011). Resource allocation schemes are proposed in the study to evaluate the system with dual-carrier networks taking advantage of both in-band and out-band relay operation modes. Simulation results showed few improvements at out-band compared to in-band, while in the case of combination of in-band and out-band operation modes the system gave the highest performance and flexibility. The weakness in the study is that the proposed multimode operation required radio spectrum allocation to each mode, thereby reducing bandwidth transmission for the network.

Most modern studies have focused on in-band mode as the main solution in LTE-A specification because this solution requires only one carrier frequency at the RN. Moreover, this is cheaper than out-band mode (Gora & Redana, 2011). RN performance is restricted by the capacity of the relay link, which carries information generated by users attached to the RN and forwards it to the BS via relay link. Although the relay link seems to be a bottleneck especially in terms of throughput, enhancing the relay link capacity yields significant gains (Bulakci, 2012b; Van Den Berg, Mandjes, & Roijers, 2006; Zhang, Hong, & Xue, 2012).

Han and Wang (2010) studied the in-band mode AF relay type of transparent relay in LTE-A networks. They analyzed the performance of the uplink transmission with realistic modeling of relay link and then scheduled the limitation of the access link. The relay link and access link are time multiplexed, which means the transmission of the relay on relay link, while the reception on the access link. By using a system-level simulator, the results indicated an improvement in relay link efficiency after load balancing between the access and relay link. The authors based the study on the scheduler algorithm of subframes transmission for both relay and access links, which was constrained by relay buffer size and transmission. The work discussed improving the relay link through scheduling the frames for the relay and access link to exploit the maximum capacity for relay link. However, the algorithm could increase the time delay through the buffering in relay and did not provide a radical and effective solution to limited capacity for the relay link.

Finally, Bulakci (2012b) gave an overview of multihop relaying concepts and discussed the relay deployment within the LTE-A. He studied the impact of relay links on multihop systems, while Mumey et al. (2011) focused on topology control for efficient communications in wireless relay networks through the use of smart antennas to enhance relay performance. They presented a new algorithm to solve the beam selection problem through a simple and fast greedy algorithm. Simulation results show that the proposed algorithms provide close-to-optimal performance. The authors focused on improving the access link through beam selection algorithm. However, they did not discuss the relay link responsible for passing the users' information to the BS via the relay node.

2.10   Moving Relay (MR)

MR has similar functionality with RN but with the difference that it provides communication to users inside the vehicle (train or bus) during their journey. MR is a new innovation to improve the throughput for vehicular users in LTE-A networks (Bulakci, 2012a). MR can be deployed flexibly to increase the throughput in wireless services for passengers on public transportations within rural areas' where RNs are not available or not economically justifiable or have a weak received signal from BSs (Bulakci, 2012a).

   MR can be installed on vehicles and connected wirelessly with the BS via relay link and with passengers via access links. As a result, MR and passengers are called group mobility (Pabst et al., 2004; Peters & Heath, 2009) as shown in Fig. 2.13. In fact, group mobility can be provided anywhere a large number of users are moving together while using cellular network services. MR provides reliable services, with the assumption that the relay link has a much better channel than regular UEs (Ding et al., 2011) because the MR antennas are high and therefore have fewer obstacles in the path of radio waves in comparison to regular UE antennas (Jing & Hassibi, 2005). Moving relays are connected to an external power source via a battery charger or have their own power supply unit, which allows them to have a relatively high access to processing capabilities and to higher transmission powers. Using MRs in cellular systems is still under discussion in the 3GPP-LTE (Sui et al., 2012b). Studies have shown that through deployment of symmetrical and cooperative relays on top of trains, the QoS of a UE inside a vehicle can be significantly improved (Farber, 2011; Van Phan, Horneman, Yu, & Vihriala, 2010).

   The most promising MR such as relay on trains has a high passenger capacity. Many of these passengers use mobile services and some even use mobile broadband. Each car in the train can be installed with its individual MR, which can be

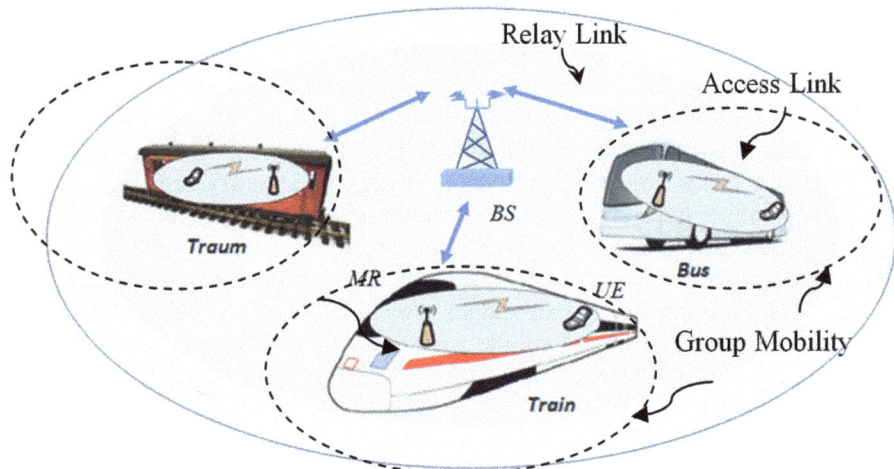

Fig. 2.13 Scenarios of moving relay (MR) installed on public transportations

interconnected. The number of MRs is based on the relay capability infrastructure (Bulakci, 2012a).

Previous studies have shown that vehicle penetration loss (VPL) can be high for UE inside a minibus at a frequency of 2.4 GHz (Tanghe, Joseph, Verloock, & Martens, 2008). The VPLs worsen with higher frequency, e.g., in LTE-A networks frequency band, for some well isolated vehicles. Therefore, the deployment of MRs on top of vehicles is very beneficial as it can eliminate VPL (Sui et al., 2012a). Most studies have been focused on fixed relay problems and handover strategies for moving relay. However, few studies have been concerned with the enhancement of throughput in public transportation users using moving relay (Van Phan et al., 2010).

Sui et al. (2012a) and Sui, Papadogiannis, Yang, and Svensson (2012b) showed the potential of moving relays, mounted on top of vehicles, to improve throughput and coverage for users in future wireless networks. They compared the performance of dual hop for moving relay assisted transmission and dual hop transmission assisted by a RN deployed on the street based on the SNR at moving relay. According to studies, the downlink of an RN system is supported with two cells: one primary cell where the outage probability performance at the vehicular UE is investigated and one interfering cell is shown in Fig. 2.14. BSs in both cells have fixed cell radius coverage, while vehicles move along a highway. When VPL is moderate to high, moving relay assisted transmission is shown to greatly outperform transmission assisted by an RN as well as direct transmission. The BS-RN and RN-UE links are denoted as relay and access links, respectively. Numerical results in the studies showed that employing moving relay on top of a public transportation vehicle can significantly enhance the QoS of the vehicular UEs. On the other hand, moving relay transmission greatly outperforms assisted RN especially, when UE moves close to cell boundaries. It is worth mentioning that these studies have introduced the important concepts of performance analysis in terms of linking the MR system with cellular networks via the RN system.

Grieger and Fettweis (2012) introduced a new solution to improve the uplink transmission for users inside vehicles by using a multiantenna in the relay: The first is outside vehicle to connect with BS, while the second is inside the vehicle to link with the users. Simulation results showed an improvement in uplink transmission for the proposed model. Often when increasing antenna diversity, there is coverage improvement. However, this requires increasing the power feeder to the antennas and ensuring the robust isolation in order to avoid self-interference between the antennas.

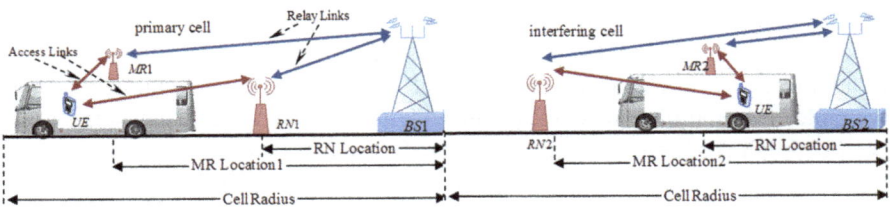

Fig. 2.14 System model of (Sui, Papadogiannis, & Svensson, 2012b)

Scott et al. (2013) discussed coverage enhancement in public transportation through deploying eight MR on board high speed train. The passengers are connected with MR via access links and MR is connected with advanced cellular networks through relay links. Simulation results showed improvements to achievable throughput of users compared to direct transmission. Although the study has improved system performance, the balance of transmission power for moving relay with received signal from the BS has not been addressed in order to minimize transmission power consumption for moving relay.

It can be seen that most studies have focused on MR analysis performances to increase users throughput inside vehicles based on channel environment and path loss. However, no author has discussed the power distribution between the relays (MR, RN) and the saving of nonrequired power for MR, especially when the vehicle passes near high resource links such as BS or RN.

Finally, there is another type of relay known as the nomadic relay node. It is a special RN case, as shown in Fig. 2.8, which is semistatic in nature. It can be deployed anywhere on the cell area such as moving relay but with the exception that while it is operational, it does not move (Frederiksen, 2008). Possible cases for using a nomadic RN could be in accident or natural disaster areas, where a sudden increase in capacity is needed. Nomadic relay nodes could also be deployed in more predictable scenarios. They might be useful during large events like sporting events, exhibitions, conferences, or demonstrations. The interval between these events might be so large that the deployment of regular BS or even RN might not be feasible (Ullah, 2012).

MR and RN are new keys introduced by LTE-A to enhance existing cellular networks and to improve the RSS within cell edge regions, crowded areas, and rural areas when BSs are inaccessible or unable to provide adequate solutions in terms of cost. In general, MR and RN represent attractive solution to meet rapid growth in future communication services (Meko, 2012; Qian Li, Qian, & Geng, 2013).

2.11 Summary

The work presented in this book has been reviewed in this chapter through the examination of the fundamentals of cooperative relaying, which consists of two basic categories; namely, Ad hoc network and multihop relay networks. The use of these networks has been elaborated with greater emphasis placed on multihop cellular networks, which can be introduced by LTE-A to enhance 3GPP-LTE cellular networks. The types of AF, DF, MR, and RN relay are studied in detail. In addition, the performance advantages of each have been highlighted and the principle of the multihop relay technique with its advantages and disadvantages has been highlighted. Earlier attempts to expand wireless network coverage with the use of multihop relay networks have also been discussed. The improvement in coverage and throughput are the main benefits of RN at LTE-A cellular networks. However, these benefits are based on location of the relay from the BS. In addition, a literature survey of this issue has been introduced. Most current publications that have various models to harness the available opportunities for multihop relay networks motivated by improving the capacity at the cell edge have also been reviewed. By examining previous studies, it was observed that almost all of the current studies rely heavily on SNR based on classical Shannon formula irrespective of interferences among cells and resource allocation such as frequencies and powers for RNs. The opportunities, challenges, and terms related to the migration from conventional cellular networks to cellular networks that use multihop relaying have been reviewed in this chapter.

References

Abdallah Bou Saleh, Ö. B., Hämäläinen, J., Redana, S., & Raaf, B. (2012). Analysis of the impact of site planning on the performance of relay deployments. *IEEE Transactions on Vehicular Technology, 61*(7), 3139–3150.

Akyildiz, I. F., Gutierrez-Estevez, D. M., & Reyes, E. C. (2010). The evolution to 4G cellular systems: LTE-advanced. *Physical Communication, 3*(4), 217–244.

Amaldi, E., Capone, A., & Malucelli, F. (2003). Planning UMTS base station location: Optimization models with power control and algorithms. *IEEE Transactions on Wireless Communications, 2*(5), 939–952.

Bletsas, A., Shin, H., & Win, M. Z. (2007). Outage optimality of opportunistic amplify-and-forward relaying. *IEEE Communications Letters, 11*(3), 261–263.

Boccardi, F., Yu, K., & Alexiou, A. (2009). Relay-aided multiple antenna transmissions for wireless backhaul applications. *Bell Labs Technical Journal, 13*(4), 161–173.

Bulakci, O. (2012a). *Multi-hop moving relays for IMT-advanced and beyond.* arXiv preprint arXiv, Department of Communications and Networking, Aalto University :1202.0207, Helsinki.

Bulakci, O. (2012b). *On backhauling of relay enhanced networks in LTE-advanced.* arXiv preprint arXiv, Department of Communications and Networking, Aalto University: 1202.0212, Helsinki.

Bulakci, O., Redana, S., Raaf, B., & Hamalainen, J. (2011). Impact of power control optimization on the system performance of relay based LTE-advanced heterogeneous networks. *Journal of Communications and Networks, 13*(4), 345–359.

Can, B., Yomo, H., & De Carvalho, E. (2007). Link adaptation and selection method for OFDM based wireless relay networks. *Journal of Communications and Networks, 9*(2), 118–127.

Chen, G. (2012). *Rate enhancement and multi-relay selection schemes for application in wireless cooperative networks.* Loughborough, England: Loughborough University.

Chiang, M. (2005). Balancing transport and physical layers in wireless multihop networks: Jointly optimal congestion control and power control. *IEEE Journal on Selected Areas in Communications, 23*(1), 104–116.

Coletti, C., Mogensen, P., & Irmer, R. (2011a, September). *Deployment of LTE in-band relay and micro base stations in a realistic metropolitan scenario.* Paper presented at the IEEE Vehicular Technology Conference (VTC Fall), San Francisco, CA.

Coletti, C., Mogensen, P., & Irmer, R. (2011b, May). *Performance analysis of relays in LTE for a realistic suburban deployment scenario.* Paper presented at the IEEE 73rd Vehicular Technology Conference (VTC Spring), Budapest.

Cover, T., & Gamal, A. E. (1979). Capacity theorems for the relay channel. *IEEE Transactions on Information Theory, 25*(5), 572–584.

Dahlman, E., Parkvall, S., & Skold, J. (2011). *4G: LTE/LTE-advanced for mobile broadband: LTE/LTE-advanced for mobile broadband*. Chichester, England: Academic Press.

Damnjanovic, A., Montojo, J., Cho, J., Ji, H., Yang, J., & Zong, P. (2012). UE's role in LTE advanced heterogeneous networks. *IEEE Communications Magazine, 50*(2), 164–176.

De Moraes, T. M., Nisar, M. D., Gonzalez, A. A., & Seidel, E. (2012). Resource allocation in relay enhanced LTE-advanced networks. *EURASIP Journal on Wireless Communications and Networking, 2012*(1), 364.

Ding, Z., Krikidis, I., Thompson, J., & Leung, K. K. (2011). Physical layer network coding and precoding for the two-way relay channel in cellular systems. *IEEE Transactions on Signal Processing, 59*(2), 696–712.

Dinnis, A., & Thompson, J. (2007, April). *The effects of including wraparound when simulating cellular wireless systems with relaying*. Paper presented at the Vehicular Technology Conference, 2007. VTC2007-Spring. IEEE 65th, Dublin.

Dong, Y., Zhang, Y., Song, M., Teng, Y., & Man, Y. (2009, September). *Optimal relay location in OFDMA based cooperative networks*. Paper presented at the WiCom'09. Proceedings of the 5th International Conference on Wireless Communications, Networking and Mobile Computing, Piscataway, NJ.

Doppler, K., Redana, S., Schultz, K., Johansson, N., Wodczak, M., Rost, P., et al. (2007). *Assessment of relay based deployment concepts and detailed description of multi-hop capable RAN protocols as input for the concept group work [R]: IST-4-027756 WINNER II*.

Drucker, E. H. (1988, June). *Development and application of a cellular repeater*. Paper presented at the IEEE 38th Vehicular Technology Conference, Philadelphia, PA.

Duong, T. Q., & Zepernick, H. J. (2009). On the performance gain of hybrid decode-amplify-forward cooperative communications. *EURASIP Journal on Wireless Communications and Networking, 2009*, 12.

Esseling, N., Walke, B. H., & Pabst, R. (2004, September). *Performance evaluation of a fixed relay concept for next generation wireless systems*. Paper presented at the 15th IEEE International Symposium on Personal Indoor and Mobile Radio Communications, PIMRC 2004, Barcelona.

Farber, M. (2011, November). *Densification in mobile networks and the potential evolution paths of the base station*. Paper presented at the 2011 8th International Symposium on Wireless Communication Systems (ISWCS), Aachen, Germany.

Fei, L., Qinghua, L., Tao, L., & Guangxin, Y. (2007). *Impact of relay location according to SER for amplify-and-forward cooperative communications*. Paper presented at the 2007 I.E. International Workshop on Anti-counterfeiting, Security& Identification.

Frederiksen, F. B. (2008). Improving spectral capacity and wireless network coverage by cognitive radio technology and relay nodes in cellular systems. *Wireless Personal Communications, 45*(3), 355–368.

Frodigh, M., Johansson, P., & LaRSSon, P. (2000). Wireless ad hoc networking-the art of networking without a network. *Ericsson Review, 4*(4), 249.

Genc, V. (2010). *Performance analysis of transparent mode IEEE 802.16 j relay-based WiMAX systems*. University College Dublin, Dublin

Genc, V., Murphy, S., & Murphy, J. (2008, April). *Performance analysis of transparent relays in 802.16 j MMR networks*. Paper presented at the 2008 6th International Symposium on Modeling and Optimization in Mobile, Ad Hoc, Wireless Networks and Workshops, Berlin.

Genc, V., Murphy, S., & Murphy, J. (2009, April). *Analysis of transparent mode IEEE 802.16 j system performance with varying numbers of relays and associated transmit power*. Paper presented at the Wireless Communications and Networking Conference, 2009 (WCNC 2009). IEEE, Budapest.

Gora, J., & Redana, S. (2011, September). *In-band and out-band relaying configurations for dual-carrier LTE-advanced system*. Paper presented at the 2011 I.E. 22nd International Symposium on Personal Indoor and Mobile Radio Communications (PIMRC),.Toronto, ON.

Grieger, M., & Fettweis, G. (2012). Field trial results on uplink joint detection for moving relays. In *Paper presented at the IEEE 8th International Conference on Wireless and Mobile Computing, Networking and Communications (WiMob)* (p. 2012).

Guo, W., Huang, X., & Liu, Y. (2010). Dynamic relay deployment for disaster area wireless networks. *Wireless Communications and Mobile Computing, 10*(9), 1238–1252.

Guo, W., Wang, S., & Chu, X. (2013). *Capacity Expression and Power Allocation for Arbitrary Modulation and Coding Rates*. Paper presented at the IEEE Wireless Communications and Networking.

Gurrala, K. K., & Das, S. (2012). Impact of relay location on the performance of multi-relay cooperative communication. *International Journal of Computer Networks and Wireless Communications, 2*(2), 226–231.

Han, J., & Wang, H. (2010, September). *Uplink performance evaluation of wireless self-backhauling relay in LTE-advanced*. Paper presented at the 2010 6th International Conference on Wireless Communications Networking and Mobile Computing (WiCOM), Chengdu, China.

Host-Madsen, A., & Zhang, J. (2005). Capacity bounds and power allocation for wireless relay channels. *IEEE Transactions on Information Theory, 51*(6), 2020–2040.

Huang, J. H., Wang, L. C., Chang, C. J., & Su, W. S. (2010). Design of optimal relay location in two-hop cellular systems. *Wireless Networks, 16*(8), 2179–2189.

Hyytiä, E., & Virtamo, J. (2007). On traffic load distribution and load balancing in dense wireless multihop networks. *EURASIP Journal on Wireless Communications and Networking, 2007*(1), 21–21.

Iwamura, M., Takahashi, H., & Nagata, S. (2010). Relay technology in LTE-advanced. *NTT DOCOMO Technical Journal, 12*(2), 29–36.

Jeon, H. c., Jung, Y. s., Kwon, B., & Ihm, J. t. (2002). *Analysis on coverage and capacity in adoption of repeater systems in CDMA 2000*. 2002 I.E. International Conference Paper presented at the 2002 International Zurich Seminar on Broadband Communications, Access, Transmission & Networking, Zurich, Switzerland

Jing, Y., & Hassibi, B. (2005, September). *Cooperative diversity in wireless relay networks with multiple-antenna nodes*. Paper presented at the International Symposium on Information Theory, 2005. ISIT 2005, Adelaide, SA.

Joshi, G., & Karandikar, A. (2011, January). *Optimal relay placement for cellular coverage extension*. Paper presented at the 2011 National Conference on Communications (NCC), Bangalore, India.

Khakurel, S., Mehta, M., & Karandikar, A. (2012, February). *Optimal relay placement for coverage extension in LTE-A cellular systems*. Paper presented at the 2012 National Conference on Communications (NCC), Kharagpur.

Khan, F. (2009). *LTE for 4G Mobile Broadband Air Interface Technologies and Performance* (1st ed.). New York: Cambridge University Press.

Kitayama, T., Hasegawa, G., Taniguchi, Y., & Nakano, H. (2013, January). *Time slot-adding algorithm for improving bottleneck link throughput in IEEE 802.16 j relay networks*. Paper presented at the 2013 International Conference on Information Networking (ICOIN), Bangkok.

Krishnan, N., Yates, R. D., Mandayam, N. B., & Panchal, J. S. (2012). Bandwidth sharing for relaying in cellular systems. *IEEE Transactions on Wireless Communications, 11*(1), 117–129.

Kwon, B., Chang, Y., & Copeland, J. A. (2008, November). *A network entry protocol and an OFDMA symbol allocation scheme for non-transparent relay stations in IEEE 802.16 j MMR networks*. Paper presented at the Military Communications Conference, San Diego, CA.

Kyungmi, P., Hyun S, R., Chung G, K., Daeyoung, C., Seungho, S., Jongguk, A., et al. (2009). The performance of relay-enhanced cellular OFDMA-TDD network for mobile broadband wireless services. *EURASIP Journal on Wireless Communications and Networking, 2009*. Article No. 5.

Laneman, J. N., Tse, D. N., & Wornell, G. W. (2004). Cooperative diversity in wireless networks: Efficient protocols and outage behavior. *IEEE Transactions on Information Theory, 50*(12), 3062–3080.

Lee, K. H., Han, K. Y., Song, J. Y., & Cho, D. H. (2006). Capacity enhancement of uplink channel through spatial reuse in multihop cellular networks. *IEEE Communications Letters, 10*(2), 76–78.

Li, X. J., Seet, B. C., & Chong, P. H. J. (2008). Multihop cellular networks: Technology and economics. *Computer Networks, 52*(9), 1825–1837.

Lin, B., & Ho, P.-H. (2007). Dimensioning and location planning of broadband wireless networks under multi-level cooperative relaying. *IEEE Transactions on Wireless Communications, 8* (11), 5682–5691.

Lin, W., Wu, G., Zhang, L., & Li, S. (2009, March). *SER performance analysis and optimal relay location of cooperative communications with distributed Alamouti code.* Paper presented at the 43rd Annual Conference on Information Sciences and Systems, 2009 (CISS 2009), Baltimore, MD.

Liu, H., Wan, P., & Jia, X. (2006). On optimal placement of relay nodes for reliable connectivity in wireless sensor networks. *Journal of Combinatorial Optimization, 11*(2), 249–260.

Madan, R., Borran, J., Sampath, A., Bhushan, N., Khandekar, A., & Ji, T. (2010). Cell association and interference coordination in heterogeneous LTE-A cellular networks. *IEEE Journal on Selected Areas in Communications, 28*(9), 1479–1489.

Martins, A., Rodrigues, A., & Vieira, P. (2012, September). *Finding optimized positioning for fixed relay stations in a cooperative LTE network.* Paper presented at the 2012 15th International Symposium on Wireless Personal Multimedia Communications (WPMC), Taipei, Taiwan.

Meko, S. F. (2012). Optimal relay placement schemes in OFDMA cellular networks. *International Journal of Engineering Research and Applications, 2*(4), 1501–1509.

Mumey, B., Tang, J., Xing, Y., & Wolff, R. (2011, December). *Relay beam selection with directional antennas.* Paper presented at the 2011 I.E. Global Telecommunications Conference (GLOBECOM 2011), Houston, TX.

Muñoz, J., Coll-Perales, B., & Gozalvez, J. (2010, October). *Research testbed for field testing of multi-hop cellular networks using mobile relays.* Paper presented at the 2010 I.E. 35th Conference on Local Computer Networks (LCN), Denver, CO.

Ng, T. C. Y., & Yu, W. (2007). Joint optimization of relay strategies and resource allocations in cooperative cellular networks. *IEEE Journal on Selected Areas in Communications, 25*(2), 328–339.

Nordio, A., Chiasserini, C., & ElBatt, T. (2012). Fair traffic relaying for two-source-one-destination wireless networks. *IEEE Wireless Communications Letters, 1*(1), 1–4.

Pabst, R., Esseling, N., & Walke, B. H. (2005). Fixed relays for next generation wireless systems–System concept and performance evaluation. *Journal of Communications and Networks, 7*(2), 104–114.

Pabst, R., Walke, B. H., Schultz, D. C., Herhold, P., Yanikomeroglu, H., Mukherjee, S., et al. (2004). Relay-based deployment concepts for wireless and mobile broadband radio. *IEEE Communications Magazine, 42*(9), 80–89.

Peters, S. W., & Heath, R. W. (2009). The future of WiMAX: multihop relaying with IEEE 802.16 j. *IEEE Communications Magazine, 47*(1), 104–111.

Prommak, C., & Wechtaison, C. (2012). Network planning and optimization for multi-hop relay placement in WiMAX networks. *Journal of Computer Science, 8*(9), 1414.

Qian Li, R. Q. H., Qian, Y., & Geng, W. (2013). Intracell cooperation and resource allocation in a heterogeneous network with relays. *IEEE Transactions on Vehicular Technology, 62*(4), 1770–1784.

Rahman, M., & Ernstrom, P. (2004). Repeaters for hotspot capacity in DS-CDMA networks. *IEEE Transactions on Vehicular Technology, 53*(3), 626–633.

Rappaport, T. S. (1996). *Wireless communications: Principles and practice* (2). Upper Saddle River, NJ: Prentice Hall PTR.

Rizinski, M., & Kafedziski, V. (2011, October). *Achievable rates of the amplify-and-forward strategy for the Gaussian relay channel.* Paper presented at the 2011 10th International Conference on Telecommunication in Modern Satellite Cable and Broadcasting Services (TELSIKS), Niš, Serbia.

Sadek, A. K., Han, Z., & Liu, K. (2010). Distributed relay-assignment protocols for coverage expansion in cooperative wireless networks. *IEEE Transactions on Mobile Computing, 9*(4), 505–515.

Salem, M., Adinoyi, A., Rahman, M., Yanikomeroglu, H., Falconer, D., Kim, Y.-D., et al. (2010). An overview of radio resource management in relay-enhanced OFDMA-based networks. *IEEE Communications Surveys & Tutorials, 12*(3), 422–438.

Satish Kumar, D., & Nagarajan, N. (2012). A new adaptive model for throughput enhancement and optimal relay selection in IEEE 802.16j networks. *Wseas Transactions on Communications, 11* (2), 82–90.

Scott, S., Leinonen, J., Pirinen, P., Vihriala, J., Van Phan, V., & Latva-aho, M. (2013, June). *A cooperative moving relay node system deployment in a high speed train.* Paper presented at the 2013 I.E. 77th Vehicular Technology Conference (VTC Spring), Dresden, Germany.

Seo, H., Mok, I., & Lee, B. G. (2007). Determination of optimal transmission power in wireless relay networks with generalized error model. *IEEE Transactions on Wireless Communications, 6*(12), 4233–4237.

Sesia, S., Toufik, I., & Baker, M. (2011). *LTE: The UMTS long term evolution.* Wiley Online Library.

Sharma, V., & Jain, D. (2010). Multihop cellular networks: A review. *International Journal of Engineering Science, 2*(11), 6082–6091.

Sharma, S., Shi, Y., Hou, Y. T., Sherali, H. D., & Kompella, S. (2010, March). *Cooperative communications in multi-hop wireless networks: Joint flow routing and relay node assignment.* Paper presented at the Infocom, 2010, Proceedings IEEE, San Diego, CA.

Sui, Y., Papadogiannis, A., Yang, W., & Svensson, T. (2012b, December). *Performance comparison of fixed and moving relays under co-channel interference.* Paper presented at the 2012 I.E. Globecom Workshops (GC Workshops), Anaheim, CA.

Sui, Y., Papadogiannis, A., & Svensson, T. (2012a, May). *The potential of moving relays-A performance analysis.* Paper presented at the 2012 I.E. 75th Vehicular Technology Conference (VTC Spring), Yokohama, Japan.

Sui, Y., Papadogiannis, A., & Svensson, T. (2012a, May). *The potential of moving relays-A performance analysis.* Paper presented at the 2012 I.E. 75th Vehicular Technology Conference (VTC Spring), Yokohama.

Tanghe, E., Joseph, W., Verloock, L., & Martens, L. (2008). Evaluation of vehicle penetration loss at wireless communication frequencies. *IEEE Transactions on Vehicular Technology, 57*(4), 2036–2041.

Ullah, I. (2012). *Performance Analysis of LTE-Advanced Relay Node in Public Safety Communication.* Finland: Aalto University.

Van Den Berg, H., Mandjes, M., & Roijers, F. (2006). Performance modeling of a bottleneck node in an IEEE 802.11 ad-hoc network. In *Ad-Hoc, mobile, and wireless networks* (pp. 321–336): New York: Springer.

Van Der Meulen, E. C. (1971). Three-terminal communication channels. *Advances in Applied Probability, 3*(1), 120–154.

Van Phan, V., Horneman, K., Yu, L., & Vihriala, J. (2010, December). *Providing enhanced cellular coverage in public transportation with smart relay systems.* Paper presented at the 2010 I.E. Vehicular Networking Conference (VNC), Jersey City, NJ.

Vidal, J., Marina, N., & Host-Madsen, A. (2008, June). *Dimensioning cellular networks with fixed relay stations.* Paper presented at the International Conference on Telecommunications, 2008 (ICT 2008), St. Petersburg, Russia.

Viswanathan, H., & Mukherjee, S. (2005). Performance of cellular networks with relays and centralized scheduling. *IEEE Transactions on Wireless Communications, 4*(5), 2318–2328.

Wang, L. C., Su, W. S., Huang, J. H., Chen, A., & Chang, C. J. (2008, March–April). *Optimal relay location in multi-hop cellular systems.* Paper presented at the 2008. WCNC 2008. IEEE Wireless Communications and Networking Conference, Las Vegas, NV.

Wang, H., Wang, J., & Xu, J. (2010). *A method and apparatus for load balance in a relay-based multi-hop wireless network.* Patent number WO 2,010,115,463, Applicant Nokia Siemens Networks OY.

Wei, H. Y., & Gitlin, R. D. (2004). Two-hop-relay architecture for next-generation WWAN/WLAN integration. *IEEE Wireless Communications Magazine, 11*(2), 24–30.

Williams, B., & Camp, T. (2002, March). *Comparison of broadcasting techniques for mobile ad hoc networks.* Paper presented at the 3rd ACM international symposium on Mobile ad hoc networking & computing, New York, NY.

Wirth, T., Thiele, L., Haustein, T., Braz, O., & Stefanik, J. (2010, September). *LTE amplify and forward relaying for indoor coverage extension.* Paper presented at the 2010 I.E. 72nd Vehicular Technology Conference Fall (VTC 2010-Fall), Ottawa, ON.

Wyglinski, A. M., Nekovee, M., & Hou, T. (2010). *Cognitive radio communications and networks: Principles and practice.* Chichester, England: Academic Press.

Xu, W., Dong, X., & Lu, W.-S. (2011). Joint precoding optimization for multiuser multi-antenna relaying downlinks using quadratic programming. *IEEE Transactions on Communications, 59* (5), 1228–1235.

Yang, Y., Hu, H., Xu, J., & Mao, G. (2009). Relay technologies for WiMAX and LTE-advanced mobile systems. *IEEE Communications Magazine, 47*(10), 100–105.

Yongchul, K., & Sichitiu, M. L. (2011). Optimal placement of transparent relay stations in 802.16 j mobile multihop relay networks. *IEICE Transactions on Communications, 94*(9), 2582–2591.

Yu, Y., Murphy, S., & Murphy, L. (2008). *Planning base station and relay station locations in IEEE 802.16 j multi-hop relay networks.* Paper presented at the 5th IEEE Consumer Communications and Networking Conference, CCNC 2008, Dublin, Ireland.

Zeng, H., & Zhu, C. (2008, September). *System-Level modeling and performance evaluation of multi-hop 802.16 j systems.* Paper presented at the International Wireless Communications and Mobile Computing Conference, 2008. IWCMC'08, College Park, MD.

Zhang, H., Hong, P., & Xue, K. (2012, January). *Uplink performance of LTE-based multi-hop cellular network with out-of-band relaying.* Paper presented at the 2012 I.E. Consumer Communications and Networking Conference (CCNC), Las Vegas, NV.

Zhao, J., Hammerstroem, I., Kuhn, M., Wittneben, A., Herdin, M., & Bauch, G. (2007). *Coverage analysis for cellular systems with multiple antennas using decode-and-forward relays.* Paper presented at the Vehicular Technology Conference, 2007. VTC2007-Spring. IEEE 65th.

Zhao, Y., Fang, X., Huang, R., & Fang, Y. (2014). Joint interference coordination and load balancing for OFDMA multi-hop cellular networks. *IEEE Transactions on Mobile Computing, 13*(1), 11.

Zheng, K., Lei, L., Wang, Y., Lin, Y., & Wang, W. (2011). Quality-of-service performance bounds in wireless multi-hop relaying networks. *IET Communications, 5*(1), 71–78.

Chapter 3
Capacity and Coverage Analysis for Multi-Hop Relay in LTE-A Cellular Network

3.1 Introduction

This chapter introduces three new methodologies to enhance the performance of LTE cellular networks (Yahya, Aldhaibani, & Ahmed, 2014). The first methodology focuses on RN deployment in the cell and called Optimum RN Deployment (ORND) to enhance the coverage area and capacity at cell-edge region. RN is considered as a solution to address low SINR at the cell edge, resolve coverage holes due to shadowing, and to meet the access requirement of nonuniform distributed traffic in densely populated areas to improve coverage and throughput. However, the interference between stations is an important problem that is associated with the RN deployment in the cell. This methodology considers the mitigation of the interferences between the stations and ensures the best capacity with the optimization of transmission power. ORND is based on mathematical analysis of determination of the optimum location for RN, optimal number of relays per cell, suitable power for each RN, and design a frequency reuse scheme, which exploits available radio spectrum. Second methodology, called Enhance Relay Link Capacity (ERLC), addresses the relay link problem, where this link carries information generated by the RN and users attached to it to BS. Although the long distance between the proposed relay location and BS improves the coverage at cell boundaries, this distance also degrades the relay link efficiency and increases the probability of outage. On the other hand, the approximation of the relay location does not achieve the desired goals to enhance coverage at the cell-edge region. ERLC introduces active solution and easy implementation to solve relay link problem. The values of parameters which are used in these models are based on the LTE system specifications and presented by (3GPP, TS. ETSI, 2007) and mentioned in Table 4.1. Third methodology focuses on enhancing the throughput and RSS for the users inside public transportation vehicles as well as proposes a new algorithm called Balance Power Algorithm (BPA) that aims to minimize the transmission power consumption for Moving Relay (MR). Summary of research on improving coverage and capacity is shown in Fig. 3.1.

© Springer International Publishing Switzerland 2017   41
A. Yahya, *LTE-A Cellular Networks*, DOI 10.1007/978-3-319-43304-2_3

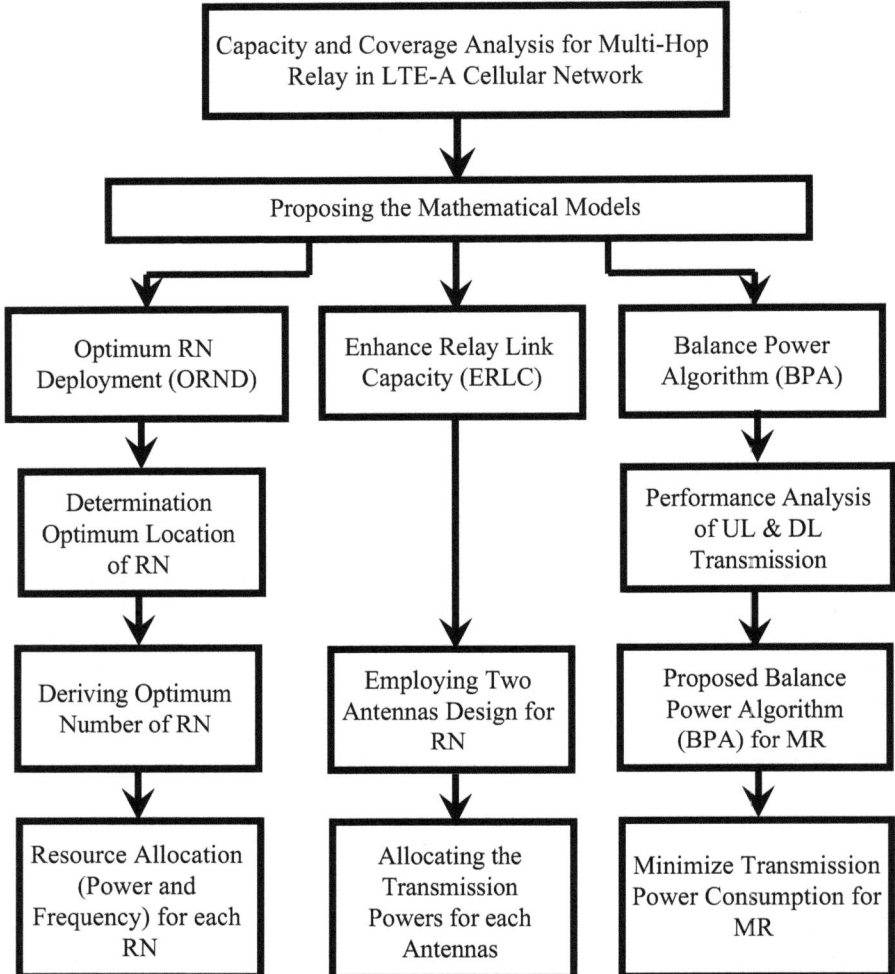

Fig. 3.1 Scope of this chapter

3.2 Channel Interference

Current researches works focus on improvement of cell-edge performance and optimize the capacity. In general, the SINR varies between cell center and cell edge at cellular networks. Thus, the capacity for users in the cell edge is lower than cell center due to a large discrepancy in QoS (Khan, 2009). Cell edge is defined as the region in which the interference power from neighboring cells is similar or stronger than the signal power from the donor BS (Khan, 2009). Although increasing the transmit power for BSs improves SINR at cell edge but also increases the interference power between cells as well as for small cells, the intercell interference (ICI) degrades cell edge SINR (Sesia, Toufik, & Baker, 2011).

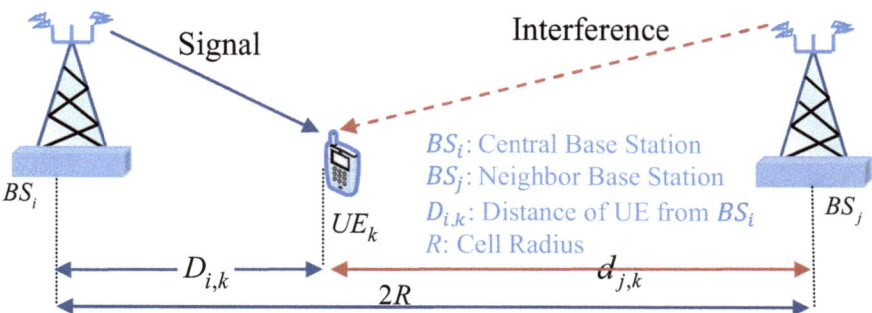

Fig. 3.2 Intercell interference scheme at UE from neighboring cell

Due to the assumption of focusing. This book on LTE system refers to a base station (BS) by the 3GPP-LTE and sometimes called Node B (eNB) (3GPP, TS. ETSI, 2007). In Fig. 3.2, when UE moves away from the cell center, SINR degrades due to two factors. Firstly, the RSS goes down because the path loss increases with distance from the BS_i. Secondly, the ICI increases because the UE moves away from one BS_i and approaching to another BS_j. UE connected to BS_i and moves away towards BS_j; therefore, the transmitted signal from BS_j appears as interference to the UE as shown in Fig. 3.2.

The DL received signal for user k can be represented as the following equation (Sadek, Han, & Liu, 2010):

$$Y_{i,k} = \sqrt{P_i} H_{i,k} X_{i,k} + \sum_{j=0}^{N_{\text{cell}}} \sqrt{P_j} H_{j,k} X_{j,k} + N_k \tag{3.1}$$

where $j = 0 \rightarrow N_{\text{cell}}$; N_{cell} is the number of neighboring cells; P_i and P_j are the transmit power of donor BS_i and neighboring BS_j, respectively; $H_{i,k}$ and $H_{j,k}$ are the fading channel gain for donor and neighboring cell, respectively. $X_{j,k}$ and $X_{j,k}$ are received signals from donor and neighboring BSs, respectively, and N_k is Additive white Gaussian noise (AWGN) for user k (Kosta, Hunt, Quddus, & Tafazolli, 2013; Mei, Bigham, Jiang, & Bodanese, 2013; Sesia et al., 2011). So the SINR at user k is

$$\rho_{i,k} = \frac{P_i |H_{i,k}|^2}{N_k + \sum_{j=0}^{N_{\text{cell}}} P_j |H_{j,k}|^2} \tag{3.2}$$

where $\rho_{i,k}$ is the SINR from the BS_i to k-UE. The channel gain H is the function of path loss (Sadek et al., 2010; Sesia et al., 2011), therefore

$$|H|^2 = D^{-\alpha} \tag{3.3}$$

Then

$$\rho_{i,k} = \frac{P_i D_{i,k}^{-\alpha}}{N_k + \sum_{j=1}^{N} P_j d_{j,k}^{-\alpha}} \tag{3.4}$$

The background noise N_k can be ignored to simplify the derivation (Khan, 2009; Sadek et al., 2010); therefore, the following is obtained

$$\rho_{i,k} = \frac{P_i D_{i,k}^{-\alpha}}{\sum_{j=1}^{N} P_j d_{j,k}^{-\alpha}} \tag{3.5}$$

where α is the path loss exponent, and $D_{i,k}$ and $d_{j,k}$ are the distances from UE to BS$_i$ and BS$_j$, respectively.

3.3   Network Capacity Without RN

The use of Adaptive Modulation and Coding (AMC) is one of the basic enabling techniques in the standards for 3G wireless networks that have been developed to achieve high spectral efficiency on fading channels (Korowajczuk, 2011). Typically, the quality of the signal received by a UE depends on channel quality from BS, level of interference from neighboring cells, and noise level. To improve system capacity and coverage for a given transmission power, the transmitter must match the bit rate for each user to the changes in the received signal (Sesia et al., 2011). This is commonly referred to as link adaptation and is typically based on AMC (Kitayama, Hasegawa, Taniguchi, & Nakano, 2013).

For a given modulation, the code rate can be chosen depending on the radio link conditions. Therefore, at DL data transmissions in LTE, the BS usually selects the code of modulation scheme according to the Channel Quality Indicator (CQI) feedback transmitted by the UE in the uplink (Song et al., 2006). This technique splits the DL capacity from BS in to two regions according to Modulation and Coding Scheme (MCS): one region is known as saturation capacity located from BS to saturation distance (X_s), while the second locates from X_s to cell radius (R) as shown in Fig. 3.3.

The classical Shannon formula is valid for infinite delay, and it is unrealistic transmission with LTE system because LTE system used MIMO and adaptive MCS technologies (Mogensen et al., 2007) as explained in Appendix A. In order to facilitate accurate benchmarking of LTE, Shannon capacity bound is modified by Wang, Wang, Xu, Teng, and Horneman (2011), Wang, Wang, and Xu (2010), and Wang et al. (2009) Thus, the capacity for LTE system in a single-input single-output (SISO) can be estimated by

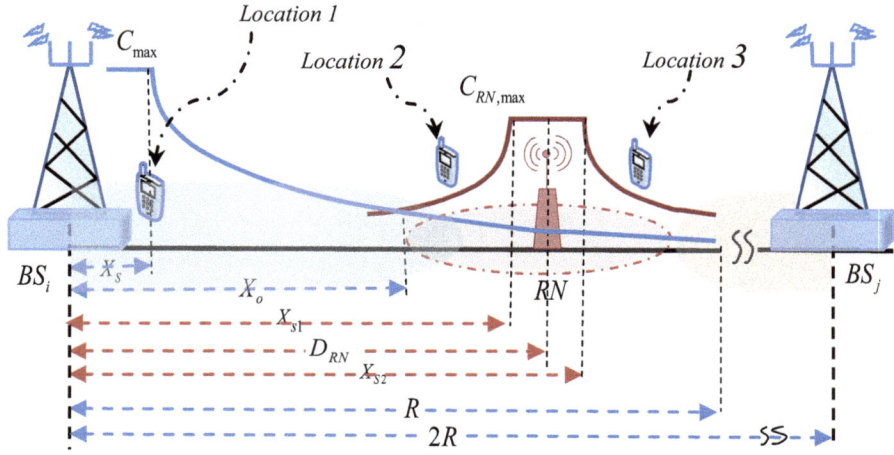

C: Spectral efficiency of node.
X_s: Distance of saturation capacity at BS_i.
X_o: Handover distance.
X_{s1}: Distance of saturation capacity at RN (left side).
X_{s2}: Distance of saturation capacity at RN (right side).
D_{RN}: RN location.
R: Cell radius.
1, 2, 3: Locations of UEs.

Fig. 3.3 Proposed model of network capacity calculation for ORND

$$\left\{ \begin{array}{ll} C_{\max} & 0 < D_i \leq X_S \\ BW_{\mathrm{eff}}\log_2(1 + \rho_i/\rho_{\mathrm{eff}}) & X_S < D_i \leq R \end{array} \right\} \qquad (3.6)$$

where C_i is the estimated spectral efficiency in bps/Hz (Wang et al., 2010), and $C_{\max}$ is the upper limit based on the hard spectral efficiency given by 64-QAM with the coding rate of 0.753 equal to 4.32 bps/Hz (Almutairi & Salamah, 2006) as presented in Appendix A. ρ_i is the SINR for each user in the cell, BW_{eff} is the adjustment for the system bandwidth efficiency, and ρ_{eff} is the adjustment for SINR implementation efficiency (Wang et al., 2009, 2011).

(BW_{eff}, ρ_{eff}) have the values of (0.56, 2.0) in the downlink and (0.52, 2.34) in the uplink (Wang et al., 2009). The proposed system used two regions in cell capacity distribution based on adaptive modulation and coding scheme. The first region around the BS is known as the estimated saturation capacity, which is specified from $0 \rightarrow X_s$ while the other region is determined from $X_s \rightarrow R$ so that in this region the cell capacity is not constant based on the modified Shannon formula distribution, as demonstrated in Fig. 3.3. According to these two regions, the received signal for UE in saturating capacity distance (X_s), as shown in Fig. 3.3, can be written as

$$Y_{i,x_s} = \sqrt{P_i} H_{i,x_s} X_{i,x_s} + \sum_{j=0}^{N_{cell}} \sqrt{P_j} H_{j,x_s} X_{j,x_s} + N_{x_s} \tag{3.7}$$

The ideal SINR at X_s location is (Wang et al., 2009)

$$\rho_{\text{ideal}} = \frac{P_i X_s^{-\alpha}}{P_j (2R - X_s)^{-\alpha}} \tag{3.8}$$

The Error Vector Magnitude (EVM) is a measure of the difference between the ideal symbols and the measured symbols after the equalization (Sesia et al., 2011). For 64-QAM modulation in LTE, the SINR (ρ_{i,X_s}) at X_S location is explained as (Sesia et al., 2011; Wang et al., 2010)

$$\frac{1}{\rho_{i,X_S}} = \frac{1}{\frac{1}{\rho_{\text{max}}} + \frac{1}{\rho_{\text{ideal}}}} \tag{3.9}$$

where

$$\rho_{\text{ideal}} = \frac{P_i X_s^{-\alpha}}{P_j (2R - X_s)^{-\alpha}} \tag{3.10}$$

$$\rho_{i,X_S} = \frac{\rho_{\text{ideal}} \rho_{\text{max}}}{\rho_{\text{ideal}} + \rho_{\text{max}}} = \frac{\rho_{\text{max}} \frac{P_i X_s^{-\alpha}}{P_j (2R - X_s)}}{\rho_{\text{max}} + \frac{P_i X_s^{-\alpha}}{P_j (2R - X_s)^{-\alpha}}}$$

$$\rho_{i,X_S} = \frac{\frac{\rho_{\text{max}} P_i X_s^{-\alpha}}{P_j (2R - X_s)}}{\frac{\rho_{\text{max}} P_j (2R - X_s)^{-\alpha} + P_i X_s^{-\alpha}}{P_j (2R - X_s)^{-\alpha}}} = \frac{\rho_{\text{max}} P_i X_s^{-\alpha}}{\rho_{\text{max}} P_j (2R - X_s)^{-\alpha} + P_i X_s^{-\alpha}}$$

To find the saturation distance X_S as function for ρ_{i,X_S}

$$(2R - X_S)\left(\rho_{i,X_S} \rho_{\text{max}} P_j\right)^{-1/\alpha} + X_s \left(\rho_{i,X_S} P_i\right)^{-1/\alpha} = X_S \left(\rho_{\text{max}} P_i\right)^{-1/\alpha}$$

Hence,

$$X_S = \frac{2R\left(\rho_{i,X_S} \rho_{\text{max}} P_j\right)^{-1/\alpha}}{\left(\rho_{\text{max}} P_i\right)^{-1/\alpha} + \left(\rho_{i,X_S} \rho_{\text{max}} P_j\right)^{-1/\alpha} - \left(\rho_{i,X_S} P_i\right)^{-1/\alpha}}$$

$$X_S = \frac{2R}{1 + \left(\frac{P_i}{P_j}\right)^{-1/\alpha} \left(\rho_{i,X_S}\right)^{1/\alpha} - \left(\rho_{\text{max}}\right)^{1/\alpha}} \tag{3.11}$$

For DL LTE system, the $\rho_{max} = 0.08$ (maximum limitation on the received SINR by using EVM) with 64-QAM (3GPP, TS. ETSI, 2007; Sesia et al., 2011; Wang et al., 2009). According to Eq. (3.12), the saturation capacity distance X_s depends on the characteristics of the sender BS_i, interference from neighboring cells, and cell radius (R).

If often all BSs have the similar characteristics for the same network (Yang, Aydin, Zhang, & Maple, 2007; Yu, Murphy, & Murphy, 2008), then the distance of capacity saturation region is

$$X_s = \frac{2R}{1 + (\rho_{i,X_s})^{1/\alpha} - (\rho_{max})^{1/\alpha}} \tag{3.12}$$

$$C_{X_s} = BW_{eff} \log_2 \left(1 + \frac{\rho_{i,X_s}}{\rho_{eff}}\right) \tag{3.13}$$

From Eq. (3.7), the total capacity over cell is equal or more than the C_{max} capacity (Wang et al., 2010), then

$$BW_{eff} \log_2 \left(1 + \frac{\rho_{i,X_s}}{\rho_{eff}}\right) \geq C_{max} \tag{3.14}$$

$$\rho_{i,X_s} \geq \rho_{eff} \left(\ln 2 \left(\frac{C_{max}}{BW_{eff}}\right) - 1\right) \tag{3.15}$$

$$\rho_{i,X_s} \geq \rho_{eff} \left(2^{\left(\frac{C_{max}}{BW_{eff}}\right)} - 1\right) \tag{3.16}$$

By substituting Eq. (3.17) in Eq. (3.13), the saturation distance X_s is

$$X_s \leq \frac{2R}{1 + \left(\rho_{eff} \left(2^{\left(\frac{C_{max}}{BW_{eff}}\right)} - 1\right)\right)^{1/\alpha} - (\rho_{max})^{1/\alpha}} \tag{3.17}$$

3.4 Handover Process Analysis

If the cellular network defines the status of channel availability, then the network decides the handover case and selects which channel and which cell to implement this process (Laiho, Wacker, & Novosad, 2006; Tripathi, Reed, & VanLandinoham, 1998). Handover process in multi-hop networks is important and difficult in comparison with conventional cellular network where it is necessary to be implemented reliably without disruption to any wireless service (Ekiz, Salih,

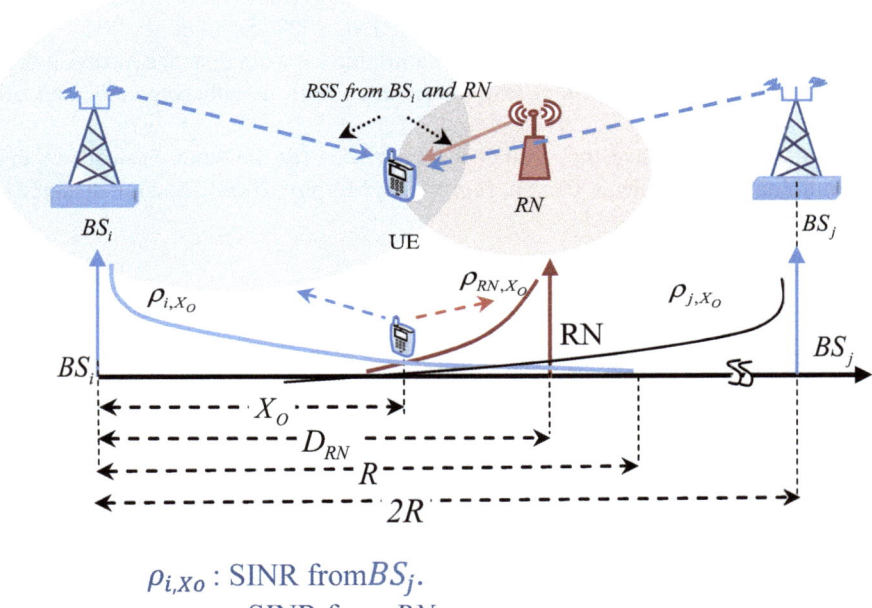

$\rho_{i,Xo}$: SINR from BS_j.

$\rho_{RN,Xo}$: SINR from RN.

Fig. 3.4 Handover process of UE movement between BS and RN

Kucukoner, & Fidanboylu, 2005). In Fig. 3.4, RSS from donor BS_i along with RSSs from neighboring BS_j must be known in order to analyze the handover process; thus, UE chooses the strongest in terms of RSS for accessing.

For multi-hop networks, RN and BS are located at a certain distance from each other; therefore handover appears when the SINR at UE from the RN is equal to the SINR from BS (Cho, Jang, & Cioffi, 2009; Kacerginskis & Narbutaite, 2012). Thus, the distance from the BS to the handover location is denoted X_o, as demonstrated in Fig. 3.4. To determine the handover distance (X_o), the received signals from both BS and RN to UE in handover location should be evaluated as follows:

The distance from relay location (D_{RN}) to X_o is $D_{RN} - X_o$; thus, the received signal from the BS at the UE in X_o point can be expressed as

$$Y_{i,X_o} = \sqrt{P_i H_{i,X_o}} X_{i,X_o} + \sqrt{P_{RN}} H_{RN,X_o} X_{RN,X_o} + N_{X_o} \qquad (3.18)$$

SINR at X_o through direct link is

$$\rho_{i,X_o} = \frac{P_i \left| H_{i,X_o} \right|^2}{P_{RN} \left| H_{RN,X_o} \right|^2} \qquad (3.19)$$

$$\rho_{i,X_o} = \frac{P_i X_o^{-\alpha}}{P_{RN}(D_{RN} - X_o)^{-\alpha}} \qquad (3.20)$$

Received signal from RN can be expressed as

$$Y_{RN,X_o} = \sqrt{P_{RN}} H_{RN,X_o} X_{RN,X_o} + \sqrt{P_i} H_{i,X_o} X_{i,X_o} + N_{X_o} \qquad (3.21)$$

$$\rho_{RN,X_o} = \frac{P_{RN} |H_{RN,X_o}|^2}{P_i |H_{i,X_o}|^2} \qquad (3.22)$$

$$\rho_{RN,X_o} = \frac{P_{RN}(D_{RN} - X_o)^{-\alpha}}{P_i X_o^{-\alpha}} \qquad (3.23)$$

At X_o, SINR from BS_i equals SINR from RN, as shown in Fig. 3.4, and based on Eqs. (3.23) and (3.24), therefore X_o can be derived as

$$\frac{P_i X_o^{-\alpha}}{P_{RN}(D_{RN} - X_o)^{-\alpha}} = \frac{P_{RN}(D_{RN} - X_o)^{-\alpha}}{P_i X_o^{-\alpha}} \qquad (3.24)$$

$$(P_i X_o^{-\alpha})(P_i X_o^{-\alpha}) = (P_{RN}(D_{RN} - X_o)^{-\alpha})(P_{RN}(D_{RN} - X_o)^{-\alpha})$$

$$P_i^{-1/\alpha} X_o = P_{RN}^{-1/\alpha}(D_{RN} - X_o)$$

$$P_{RN}^{-1/\alpha} D_{RN} = P_i^{-1/\alpha} X_o + P_{RN}^{-1/\alpha} X_o$$

$$P_{RN}^{-1/\alpha} D_{RN} = X_o \left(P_i^{-1/\alpha} + P_{RN}^{-1/\alpha} \right)$$

$$X_o = \frac{P_{RN}^{-1/\alpha} D_{RN}}{\left(P_i^{-1/\alpha} + P_{RN}^{-1/\alpha} \right)} = \frac{D_{RN}}{\left(\frac{P_i^{-1/\alpha}}{P_{RN}^{-1/\alpha}} + 1 \right)}$$

$$X_o = \frac{D_{RN}}{\left(\left(\frac{P_{RN}}{P_i} \right)^{1/\alpha} + 1 \right)} \qquad (3.25)$$

X_o is a distance that depends on the relay location, node characteristics, and path loss exponent (α).

3.5   Network Capacity with RN

Link adaptation process of relay networks is similar to that for the BS when using DF relay, since the choice of modulation scheme to use depends on the current state of the transmission channel (3GPP, ETSI. T., 2009; Sesia et al., 2011) as shown in Fig. 3.5.

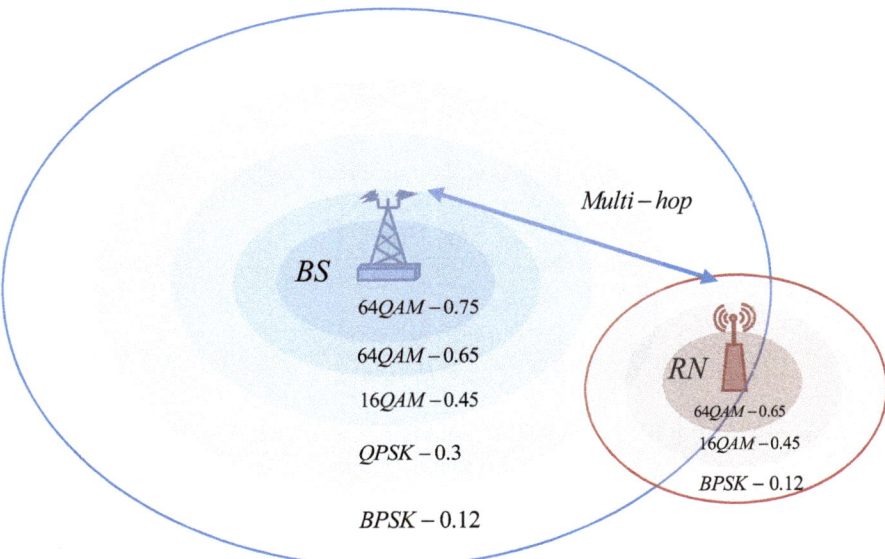

Fig. 3.5 Transmission range of the modulation and coding schemes in LTE cellular network (Huang et al., 2010)

Locations 1 and 2, as shown in Fig. 3.3, for UE have been chosen in order to analyze the DL capacity for RN. The received signals at UE in Locations 2 and Locations 3, respectively, as illustrated in Fig. 3.3, can be represented as

$$Y_{RN,2} = \sqrt{P_{RN}}H_{RN,2}X_{RN,2} + \sqrt{P_i}H_{i,2}X_{i,2} + N_2 \qquad (3.26)$$

$$Y_{RN,3} = \sqrt{P_{RN}}H_{RN,3}X_{RN,3} + \sqrt{P_i}H_{i,3}X_{i,3} + N_3 \qquad (3.27)$$

SINR for Locations 2 and 3 in Fig. 3.3 is

$$\rho_{RN,2} = \frac{P_{RN}|H_{RN,2}|^2}{P_i|H_{i,2}|^2} \qquad (3.28)$$

$$\rho_{RN,3} = \frac{P_{RN}|H_{RN,3}|^2}{P_i|H_{i,3}|^2} \qquad (3.29)$$

By using Eq. (3.3), SINRs for Locations 2 and 3, respectively, are

$$\rho_{RN,2} = \frac{P_{RN}(D_{RN} - D_i)^{-\alpha}}{P_i D_i^{-\alpha}} \qquad (3.30)$$

$$\rho_{RN,3} = \frac{P_{RN}(D_i - D_{RN})^{-\alpha}}{P_i D_i^{-\alpha}} \qquad (3.31)$$

Based on AMC, the capacity of relaying system is divided into four regions (i.e., two sides of DL capacity for RN) to describe the coverage distribution over cell edge according to location for RN (D_{RN}) as shown in Fig. 3.3. Therefore, the mathematical analysis is represented by the following equations:

$$
C_{\mathrm{RN},2} =
\begin{cases}
\mathrm{BW}_{\mathrm{eff}}\log_2\left(1 + \dfrac{P_{\mathrm{RN}}(D_{\mathrm{RN}} - D_i)^{-\alpha}}{P_i\rho_{\mathrm{eff}}D_i^{-\alpha}}\right) & X_{\mathrm{o}} < D_i < X_{s1} \\[3mm]
C_{R\max} & X_{s1} < D_i < D_{\mathrm{RN}}
\end{cases}
\tag{3.32}
$$

$$
C_{\mathrm{RN},3} =
\begin{cases}
C_{R\max} & D_{\mathrm{RN}} < D_i < X_{s2} \\[3mm]
\mathrm{BW}_{\mathrm{eff}}\log_2\left(1 + \dfrac{P_{\mathrm{RN}}(D_i - D_{\mathrm{RN}})^{-\alpha}}{P_i\rho_{\mathrm{eff}}D_i^{-\alpha}}\right) & X_{s2} < D_i < R
\end{cases}
\tag{3.33}
$$

$C_{R\max}$ is the upper limit based on the hard spectral efficiency given by 64-QAM with the coding rate from (0.55 to 0.65) (Almutairi & Salamah, 2006).

In order to proceed the mathematical analysis of $C_{\mathrm{RN},2}$ and $C_{\mathrm{RN},3}$, the domains X_{s1} and X_{s2}, as shown in Fig. 3.3, should be derived as follows:

Based on Eq. (3.3), SINR at X_{s1} from RN is

$$
\rho_{\mathrm{RN},X_{S1}} = \frac{(D_{\mathrm{RN}} - X_{S1})^{-\alpha}P_{\mathrm{RN}}}{X_{S1}^{-\alpha}P_i}
\tag{3.34}
$$

$$
P_i\rho_{\mathrm{RN},X_{S1}}X_{S1}^{-\alpha} = (D_{\mathrm{RN}} - X_{S1})^{-\alpha}P_{\mathrm{RN}}
\tag{3.35}
$$

$$
\left(P_i\rho_{\mathrm{RN},X_{S1}}\right)^{-1/\alpha}X_{S1} = (P_{\mathrm{RN}})^{-1/\alpha}D_{\mathrm{RN}} - X_{S1}(P_{\mathrm{RN}})^{-1/\alpha}
\tag{3.36}
$$

$$
\left(P_i\rho_{\mathrm{RN},X_{S1}}\right)^{-1/\alpha}X_{S1} + X_{S1}(P_{\mathrm{RN}})^{-1/\alpha} = (P_{\mathrm{RN}})^{-1/\alpha}D_{\mathrm{RN}}
\tag{3.37}
$$

$$
X_{S1}\left[\left(P_i\rho_{\mathrm{RN},X_{S1}}\right)^{-1/\alpha} + (P_{\mathrm{RN}})^{-1/\alpha}\right] = (P_{\mathrm{RN}})^{-1/\alpha}D_{\mathrm{RN}}
\tag{3.38}
$$

$$
X_{S1} = \frac{(P_{\mathrm{RN}})^{-1/\alpha}D_{\mathrm{RN}}}{\left[\left(P_i\rho_{\mathrm{RN},X_{S1}}\right)^{-1/\alpha} + (P_{\mathrm{RN}})^{-1/\alpha}\right]} = \frac{D_{\mathrm{RN}}}{\left[\dfrac{\left(P_i\rho_{\mathrm{RN},X_{S1}}\right)^{-1/\alpha}}{(P_{\mathrm{RN}})^{-1/\alpha}} + 1\right]}
\tag{3.39}
$$

$$
X_{s1} = \frac{D_{\mathrm{RN}}}{\left[\left(\dfrac{P_{\mathrm{RN}}}{P_i\rho_{\mathrm{RN},X_{S1}}}\right)^{1/\alpha} + 1\right]}
\tag{3.40}
$$

Similarly to find the X_{S2}

$$
\rho_{\mathrm{RN},X_{S2}} = \frac{(X_{S2} - D_{\mathrm{RN}})^{-\alpha}P_{\mathrm{RN}}}{X_{S2}^{-\alpha}P_i}
\tag{3.41}
$$

$$\rho_{RN,X_{S2}} X_{S2}^{-\alpha} P_i = (X_{S2} - D_{RN})^{-\alpha} P_{RN} \tag{3.42}$$

$$\left((P_{RN})^{-1/\alpha} - \left(\rho_{RN,X_{S2}} P_i \right)^{-1/\alpha} \right) X_{S2} = D_{RN} (P_{RN})^{-1/\alpha} \tag{3.43}$$

$$X_{S2} = \frac{D_{RN} (P_{RN})^{-1/\alpha}}{(P_{RN})^{-1/\alpha} - \left(\rho_{RN,X_{S2}} P_i \right)^{-1/\alpha}} \tag{3.44}$$

$$X_{S2} = \frac{D_{RN}}{\left(1 - \left(\frac{P_{RN}}{\rho_{RN,X_{S2}} P_i} \right)^{+1/\alpha} \right)} \tag{3.45}$$

3.6 Optimum RN Location (D_{RN})

In this section, the issue of the optimum location for RN in a dual-hop network over LTE-A cellular networks is addressed to improve the capacity with interference—limited between the stations.

Based on Eqs. (3.33) and (3.34), at the cell edge region, the DL capacity is determined from handover point $D_i = X_o$ to cell boundaries $D_i = R$; therefore, derivation of the optimum location for RN (D_{RN}) within cell edge region is

$$\text{BW}_{\text{eff}} \log_2 \left(1 + \frac{P_{RN} (D_{RN} - X_o)^{-\alpha}}{P_i \rho_{\text{eff}} X_o^{-\alpha}} \right) = \text{BW}_{\text{eff}} \log_2 \left(1 + \frac{P_{RN} (R - D_{RN})^{-\alpha}}{P_i \rho_{\text{eff}} R^{-\alpha}} \right) \tag{3.46}$$

$$\frac{(D_{RN} - X_o)^{-\alpha}}{X_o^{-\alpha}} = \frac{(R - D_{RN})^{-\alpha}}{R^{-\alpha}} \tag{3.47}$$

$$X_o (R - D_{RN}) = R(D_{RN} - X_o) \tag{3.48}$$

By substituting Eq. (3.26) in Eq. (3.49)

$$\frac{2 D_{RN} R}{1 + \left(\frac{P_{RN}}{P_i} \right)^{1/\alpha}} - \frac{D_{RN}^2}{1 + \left(\frac{P_{RN}}{P_i} \right)^{1/\alpha}} - R D_{RN} = 0 \tag{3.49}$$

$$\frac{2R - D_{RN}}{1 + \left(\frac{P_{RN}}{P_i} \right)^{1/\alpha}} = R \tag{3.50}$$

$$2R - D_{RN} = R \left(1 + \left(\frac{P_{RN}}{P_i} \right)^{1/\alpha} \right) \tag{3.51}$$

Then

$$D_{\text{RN}} = R\left(1 - \left(\frac{P_{\text{RN}}}{P_i}\right)^{1/\alpha}\right) \tag{3.52}$$

From Eq. (3.53), the location of RN depends on the cell radius, transmitted power for both BS and RN, and path loss exponent. These parameters define the location of RN between X_{o} and R.

3.7 Optimum Number of Relays (N_{relays})

To enhance the coverage area by avoiding the overlapping among RNs and providing best coverage with lower number of RNs. The optimum number of relays should be calculated to ensure the fair capacity distribution and avoiding the overlapping between neighboring RNs. Assuming that a UE is present in the midpoint between two RNs and that d_{nr} is the distance from RN to UE as illustrated in Fig. 3.6.

Although, increasing the number of RNs can improve cell capacity; however, it also increases resource allocation and overlapping between RNs. In order to avoid the overlapping, in this work, the derivation of optimum number of relays (N_{relays}) should be based on the path loss from both neighboring RNs and BS to UE. Assuming the interference power between neighboring RNs to UE be stronger than interference power from donor BS to UE, as the follow:

$$P_{\text{RN1}}(d_{\text{nr}})^{-\alpha} + P_{\text{RN2}}(d_{\text{nr}})^{-\alpha} > P_i(D_{\text{RN}})^{-\alpha} \tag{3.53}$$

and

$$2P_{\text{RN}}(d_{\text{nr}})^{-\alpha} > P_i(D_{\text{RN}})^{-\alpha} \tag{3.54}$$

As shown in Fig. 3.6, RNs deployed on the circumference of a circle around the BS with radius D_{RN}; thus, the distance between neighboring RNs is

$$2d_{\text{nr}} = \frac{2\pi D_{\text{RN}}}{N_{\text{relays}}} \tag{3.55}$$

By substituting with Eq. (3.55)

$$\frac{2P_{\text{RN}}}{P_i}\left(\frac{\pi D_{\text{RN}}}{N_{\text{relays}}}\right)^{-\alpha} > (D_{\text{RN}})^{-\alpha} \tag{3.56}$$

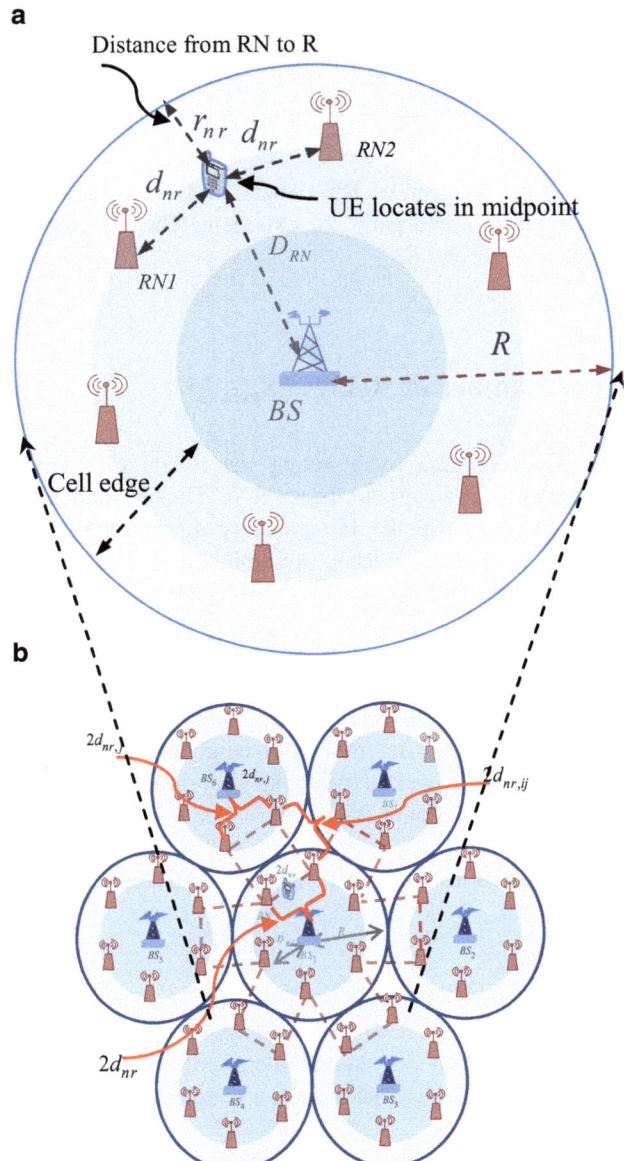

Fig. 3.6 Constraints of determination of the optimum number of RNs per cell (**a**) RN deployment for one cell (**b**) RN deployment for first tier

Thus, the optimum number of relays (N_{relays}) is

$$N_{\text{relays}} < \pi \left(\frac{2P_{\text{RN}}}{P_i} \right)^{-1/\alpha} \tag{3.57}$$

To mitigate interference between all RNs in LTE-A cellular network, let $2d_{nr} = 2d_{nr,j} = 2d_{nri,j}$, as shown in Fig. 3.6b, where N_{relays} is optimum number of RNs in a cell, $2d_{nr,j}$ is distance between the neighboring RNs in BS$_j$, and $2d_{nri,j}$ is the distance between the neighboring RNs of first tier network, as indicated in Fig. 3.6b.

From Eq. (3.58), N_{relays} depends on the transmission power of both RN and BS as well as on the path loss exponent (α). Therefore, a higher RN transmit power means fewer RNs should be deployed in cell to minimize mutual overlapping between RNs and BS.

3.8   Pseudo-Codes of RN Deployment

This section presents pseudo-codes of RN deploying in LTE-A cellular network. The proposed ORND approach consist five pseudo-codes in order to enhance the capacity and coverage extension based on proposed mathematical model. These pseudo-codes are implemented using MATLAB R2011 software (Gibson, 2012). The five pseudo-codes are as follows:

Pseudo-Code 3.1 Pseudo-Codes of DL Capacity without RN

Requirement:
  C_{max}, R, Simulation parameters is in Table **4.1**

Ensure:
C_i /*DL Capacity without RN */
1:   BEGIN
2:   BS$_i$ = BS$_i$ (x_o, y_o)
2:      Calculate $\rho_{i,Xs}, X_o, X_s$
3:      **for** $i = 1$ to R
4:        **if** $D_i \geq X_s$
5:   $C_i = BW_{eff} \log_2 \left(1 + \rho_{i,Xs}/\rho_{eff}\right)$
6:        **else**
7:          $C_i = C_{max}$
8:      **end if**
9:      **end for**

END

Pseudo-Code 3.2 Pseudo-Codes of DL Capacity with RN

Requirement:

C_{Rmax}, R, Simulation parameters is in Table **4.1**

Ensure:

C_{RN} /*DL Capacity with RN */

1: BEGIN
2: $BS_i = BS_i (x_o, y_o)$
2: Calculate $\rho_{i,Xs1}$, $\rho_{i,Xs2}$, X_o, X_{s1}, X_{s2}
3:   **for** $i = X_o$ to R **do**
4:    **if** $D_i \geq X_o$
5:     **if** $D_i < X_{s1}$
6: $C_{RN,2} = BW_{eff} \log_2 \left(1 + \frac{P_{RN}(D_{RN} - D_i)^{-\alpha}}{P_i \rho_{eff} D_i^{-\alpha}}\right)$
7:     **else**
8:      **if** $D_i \geq X_{s2}$
9: $C_{RN,3} = BW_{eff} \log_2 \left(1 + \left({P_{RN}(D_i - D_{RN})^{-\alpha}}\Big/{P_i \rho_{eff} D_i^{-\alpha}}\right)\right)$
10:      **else**
11:       $C_{RN,2} = C_{Rmax}$
12: **end if**
13:     **end if**
14:    **end if**
15: $C_{RN} = C_{RN2} + C_{RN3}$
16:    **end for**

END

Pseudo-Code 3.3 Pseudo-Codes of Enhancing in DL Capacity of BS with RN

Require:

C_{Rmax}, C_{max} R, Simulation parameters is in Table **4.1**

Ensure:

$C_{W/RN}$ /* DL Capacity with RN */

1: BEGIN
2: $BS_i = BS_i (x_o, y_o)$
2: Calculate $\rho_{i,Xs}$, $\rho_{i,Xs1}$, $\rho_{i,Xs2}$, X_o, X_s, X_{s1}, X_{s2}
3: **for** $i = 0$ to R **do**
4:   **if** $D_i < X_s$

(continued)

Pseudo-Code 3.3 (continued)

5:      $C_{W/RN} = C_{\max}$
6:**else**
7:        **if $D_i < X_o$**
8:          $C_{W/RN} = BW_{eff} \log_2\left(1 + \rho_{i,X_S}/\rho_{eff}\right)$
9:**else**
10:            **if $D_i < X_{s1}$**
11:$C_{RN,2} = BW_{eff} \log_2\left(1 + \frac{P_{RN}(D_{RN}-D_i)^{-\alpha}}{P_i \rho_{eff} D_i^{-\alpha}}\right)$
12:            **else**
13:                **if $D_i \geq X_{s2}$**
14:                  $C_{W/RN} = BW_{eff} \log_2\left(1 + \left(P_{RN}(D_i-D_{RN})^{-\alpha}\middle/ P_i \rho_{eff} D_i^{-\alpha}\right)\right)$
15:                **else**
16:                  $C_{W/RN} = C_{R\max}$
17:**end if**
18:            **end if**
19:        **end if**
20:    **end if**
21: **end for**

END

Pseudo-Code 3.4: Pseudo-Codes to Determine Optimum Number of RN per Cell

Requirement:

P_{RN}, P_{BS}, R, Simulation parameters is in Table **4.1**

Ensure:
N_{relays} /*Optimal Number of RN Per Cell */
$RN(n)$ /*Distance between neighboring RN within a cell */

1: BEGIN
2: $BS_i = BS_i(x_o, y_o)$
2: Calculate N_{relays}, D_{RN}
3:    **for** $n = 1$ to N_{relays}
4:      **for** $\theta = 1$ to 360 step $(360/N_{relays})$ **do**
5:$RN(n) = (D_{RN} \sin(\theta), D_{RN} \cos(\theta))$
6:Deploy $RN(n)$
4:      **end for**
5:    **end for**

END

Pseudo-Code 3.5: Pseudo-Codes to Determine Number of Distributed Users within Cell

Require:

P_{RN}, P_{BS}, R, Q, Simulation parameters is in Table **4.1**

Ensure:

N_{RN-UE} , N_{BS-UE} /*Number of User Attached with RN and BS */

```
 1:  BEGIN
 2:  N_RN-UE = N_BS-UE = 0
 3:    for i = 1 to Q do
 4:      Calculate X_o, D_UE
 5:      if X_o ≥ D_UE then
 6:Increment N_RN-UE by 1
 7:else
 8:          Increment N_BS-UE by 1
 9:      end if
10:    end for

END
```

3.9 Frequency Reuse for Multi-Hop Relay

Multi-hop relay networks have drawn attention due to its high throughput and extensive coverage. However, the allocation of efficient frequency resource to relay links becomes a challenging design issue, especially for UEs near the cell edge because they face severe interferences from RNs within cell and from neighboring cells, which significantly affect the network performance and increase the service outage in the cell (Park et al., 2007).

For LTE cellular network, the frequency reuse pattern can be denoted as $N \times S \times K$, where the networks are divided into clusters of N cells (each cluster has a different frequency band), with S sectors and K different carrier frequencies per cell (Zhao et al., 2014) as shown in Fig. 3.7, where the pattern can be denoted as $(1 \times 3 \times 3)$. In multi-hop relay, the deployment of RNs at the cell boundaries causes high interference between the BS and RNs and between RNs themselves (Le & Hossain, 2007). Therefore, using 1x3x3 in multi-hop relaying network reduces spectral efficiency because frequency resource has to be assigned to relay links. However, this scheme mitigates the interference between the nodes (Wang et al., 2008) as illustrated in Fig. 3.7. The proposed scheme in this book, based on frequency reuse for 1 cluster and 3 sectors used 3 frequencies $(1 \times 3 \times 3)$ at BS region, so that RNs used same frequencies for BS but at different regions within the cell as shown in Fig. 3.7. Transmission power of BS and RN will determine the number of RNs and their locations as discussed in Sects. 3.6 and 3.7.

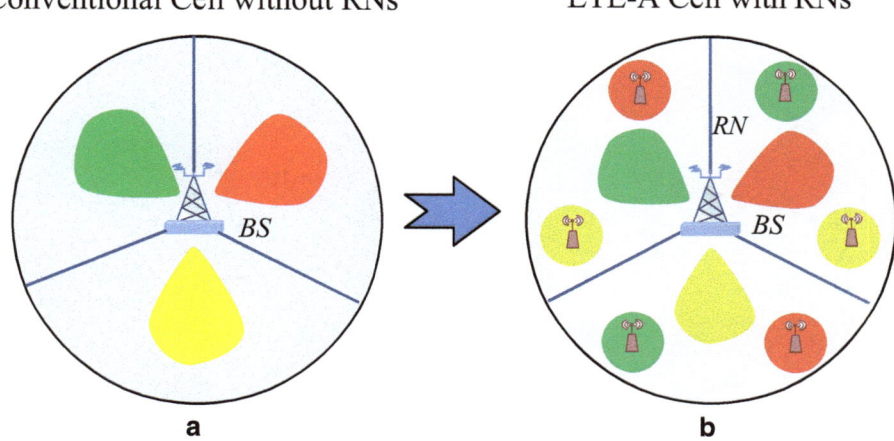

Fig. 3.7 Proposed frequency reuse for multi-Hop relay (**a**) for conventional cellular network (**b**) LTE-A cellular network using six RN deployed in the cell (6RN5W)

Figure 3.7 depicts frequency band allocation to each node; bandwidth W is divided equally into three different sub-bands: W1 (red), W2 (yellow), and W3 (green) each color indicates the sub-bands within cell. The same frequencies are reused to RNs but with different regions of RNs within cell in order to mitigate the interference between the nodes and exploit a lower available bandwidth, as shown in Fig. 3.7.

3.10   Enhance Relay Link Capacity (ERLC)

The capacity and coverage at the cell edge remain relatively small because of the low SINR, so that the capacity near the BS is better than that near the cell edge (Kosta et al., 2013; Sadek et al., 2010

RN performance is limited by the capacity of the relay link, which carries information generated by the RN and all users attached to it. Thus, a relay link is represented as a bottleneck. Although the increase in distance between the RN location and BS improves the coverage at cell boundaries, it degrades the relay link efficiency due to increasing the path loss between RN and BS. On the contrary, the approximation of the RN dose not achieve the desired goals for the enhancement of the capacity at cell boundaries.

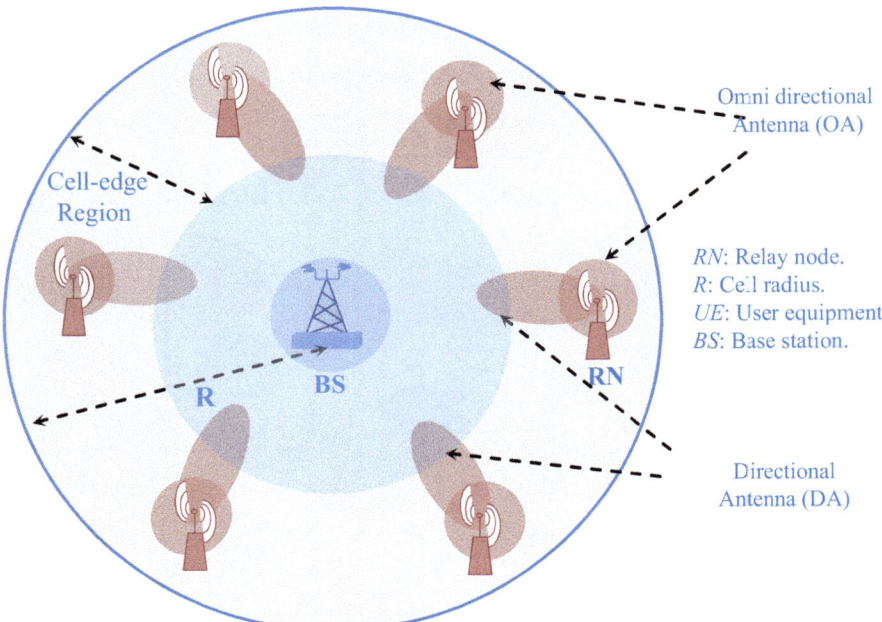

Fig. 3.8 Proposed model of using two types of antennas; OA and DA are operated as half duplex mode within six RNs deployed in cell

To solve the problem of capacity-limited for relay link, a new model is proposed which is called Enhance Relay Link Capacity (ERLC). This model employs two antennas at RN: one is directional antenna (DA) directed toward the BS in order to improve relay link and achieving flexible in the range of RN location, while the other one is omni-directional antenna (OA) to exchange information between the RN and attached users as shown in Fig. 3.8. Half duplex (HD) mode is proposed mode of operation in this scheme, where two antennas operate at same frequency with different time slots. HD mode exploits the available frequency band and avoiding the self-interference at the receiving antenna in comparison with Full Duplex (FD) mode (Chun & Park, 2012; Khafagy, Ismail, Alouini, & Aissa, 2013; Riihonen, Werner, & Wichman, 2011).

3.10.1 Handover Measurement for DA

Employing the DA affected handover location, where the handover location at UE will be closer to BS than using OA. Therefore, the enhancing of capacity for relay link in this section is derived based on evaluating handover as function of antenna gain.

Based on Eqs. (3.21) and (3.24), and the gain of antennas of BS and RN are not equal, SINR from BS at hand over location is

$$\rho_{i,X_o} = \frac{P_i G_{\mathrm{BS}} G_{\mathrm{ue}} X_o^{-\alpha}}{P_{\mathrm{RN}} G_d G_{\mathrm{ue}} (D_{\mathrm{RN}} - X_o)^{-\alpha}} \tag{3.58}$$

while received SINR from RN is

$$\rho_{\mathrm{RN},X_o} = \frac{P_{\mathrm{RN}} G_d G_{\mathrm{ue}} (D_{\mathrm{RN}} - X_o)^{-\alpha}}{P_i G_{\mathrm{BS}} G_{\mathrm{ue}} X_o^{-\alpha}} \tag{3.59}$$

At handover location $\rho_{i,X_o} = \rho_{\mathrm{RN},X_o}$

$$\frac{P_{\mathrm{RN}} G_d G_{\mathrm{ue}} (D_{\mathrm{RN}} - X_o)^{-\alpha}}{P_i G_{\mathrm{BS}} G_{\mathrm{ue}} X_o^{-\alpha}} = \frac{P_i G_{\mathrm{BS}} G_{\mathrm{ue}} X_o^{-\alpha}}{P_{\mathrm{RN}} G_d G_{\mathrm{ue}} (D_{\mathrm{RN}} - X_o)^{-\alpha}} \tag{3.60}$$

$$X_o = \frac{D_{\mathrm{RN}}}{\left(\left(\frac{G_{d,\mathrm{RN}} P_{\mathrm{RN}}}{G_{\mathrm{BS}} P_t} \right)^{1/\alpha} + 1 \right)} \tag{3.61}$$

where G_{BS}, G_{ue}, and G_d are antenna gains of OA for BS, OA for UE, and DA for RN, respectively. From Eq. (3.62), the $X_o \propto 1/G_d$, therefore any increment in the DA gain for RN, leads to approximation in the handover distance, thus, improves the spectral efficiency of the relay link.

Equations (3.63) and (3.64) explain the enhancement in SINR and spectral efficiency by DA within the distance between the location for RN and X_o

$$\rho_{\mathrm{RN},X_i} = \frac{P_{\mathrm{RN}} G_d (D_{\mathrm{RN}} - X_i)^{-\alpha}}{P_i G_{\mathrm{BS}} X_i^{-\alpha}} \quad \text{for } X_o < X_i < D_{\mathrm{RN}} \tag{3.62}$$

Then

$$C_{X_i} = \mathrm{BW}_{\mathrm{eff}} \log_2 \left(1 + \frac{\rho_{\mathrm{RN},X_i}}{\rho_{\mathrm{eff}}} \right) \tag{3.63}$$

where the ρ_{RN,X_i}, C_{X_i} is SINR and spectral efficiency at UE along distance between the location of RN (D_{RN}) and handover location (X_o). This model provides the flexibility to approximation D_{RN} toward cell boundaries to increase the coverage meanwhile guarantee high quality of link between RN and the BS.

3.10.2 Proposed Antenna

The DA is used at the RN and is directed toward the BS to improve the link quality and accommodate simultaneously users associated through the RN. The gain or directivity of an antenna is the ratio of the radiation intensity in a given direction to

Fig. 3.9 Aperture for
directional antenna

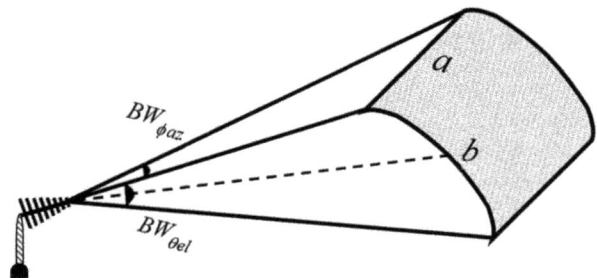

the average radiation intensity of all directions. Therefore, when the pattern of the antenna is uniform, the gain is equal to the area of the isotropic sphere $(2\pi r^2)$ divided by the sector (cross section) (Dobkin, 2012):

$$G_d = \frac{\text{Area of Sphere}}{\text{Area of Antenna Pattern}} \tag{3.64}$$

$$G_d = \frac{4\pi}{\text{BW}_{\phi az}\text{BW}_{\theta el}} = \frac{4\pi}{\sin(\phi_{az})\sin(\theta_{el})} \tag{3.65}$$

where $\text{BW}_{\phi az}$ and $\text{BW}_{\phi el}$ are the azimuth angle and beam width elevation in radians, respectively; ϕaz and ϕel are the azimuth and elevation angles, respectively, as shown in Fig. 3.9 (Dobkin, 2012). Based on Eq. (3.66), the increment in gain for the DA depends on the azimuth if the elevation angle is fixed.

A broadband directional antenna as shown in Fig. 3.10a is chosen in the simulation test as a DA with variable gain; Fig. 3.10b depicts the antenna pattern with 72° azimuth angle and directed toward BS according to (Dobkin, 2012).

3.11   System Modeling

The propagation of a radio wave is a complicated and less predictable process if the transmitter and receiver properties are considered in channel environment calculations. The process is governed by reflection, diffraction, and scattering, the intensities of which vary under different environments at different instances.

A-ATDI (Abdulrazak, Rahman, & Sharul Kamal, 2009) simulator is used to validate the mathematical models for both approaches, RN Deployment in LTE-A Cellular Network and Enhance Relay Link Capacity. Figure 3.11 shows the propagation model between the nodes (RNs and BS). This model is used by simulator described through the following equation (Korowajczuk, 2011):

$$P_r = P_t + G_t + G_r - L_{\text{prop}} - L_t - L_{\text{re}} \ [\text{dB}] \tag{3.66}$$

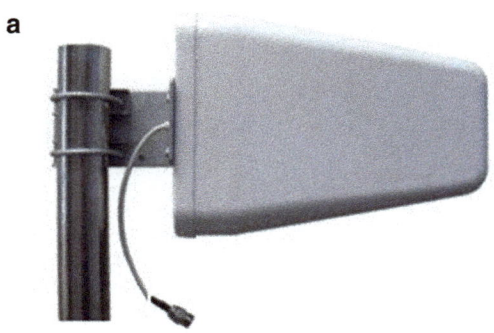

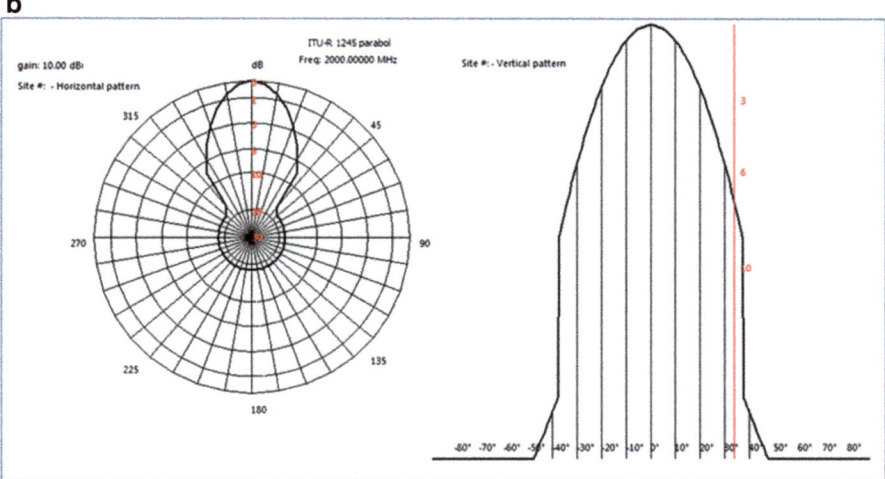

Fig. 3.10 Broadband directional antenna type (**a**) real photo (**b**) pattern simulation with gain 10 dBi

where P_t indicates the power at the transmitter and P_r is the power at the receiver; G_t and G_r are the transmitter and receiver antenna gains, respectively; L_t and L_{re} express the feeder losses; and L_{prop} is the total propagation loss, formulated as

$$L_{prop} = L_{fsd} + L_d + L_{sp} + L_{gas} + L_{rain} + L_{clut} \qquad (3.67)$$

Where
L_{fsd}: Free space distance loss,
L_d: Diffraction loss,
L_{sp}: Sub-path loss,
L_{gas}: Attenuation caused by atmospheric gas,
L_{rain}: Attenuation caused by hydrometeor scatter, and
L_{clut}: Clutter attenuation.

Equation (3.68) describes the link budget that weakens the transmitted signal before it is received by the receiver. Link budget depends on gains and losses in the

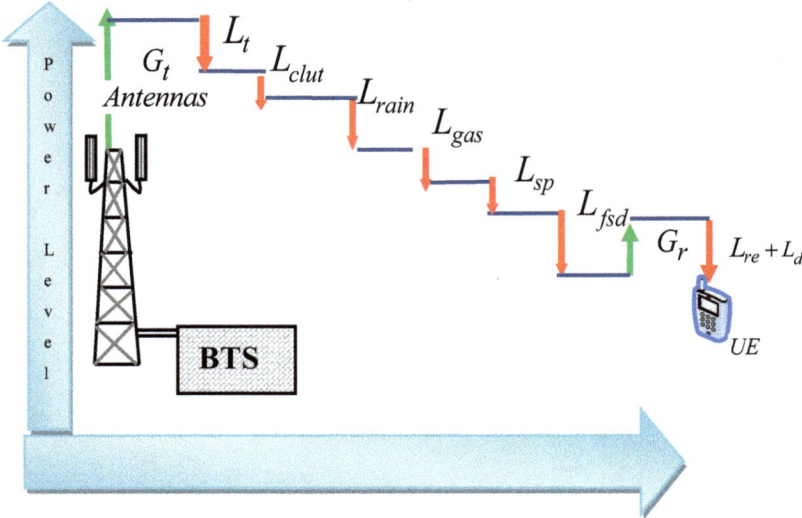

L_t: Tx. feeder loss; L_{clut}:Clutter attenuation; L_{gas}:Attenuation caused by atmospheric gas; L_{sp}: Sub-path loss; L_{rain}: Attenuation caused by hydrometeor scatter, and rain; L_{fsd}:Free space distance loss;
L_d: Diffraction loss;L_r: Rx. feeder loss; BTS: Base Transceiver Station;
G_t: Transmitter antenna gain; G_r: Receiver antenna gain

Fig. 3.11 Link budget scheme for simulation test

path, which is facing the transmitted signal to reach the receiver. A link is created by three related communication entities: transmitter, receiver, and a channel (medium) between them. The medium introduces losses caused by absorption of the transmitted power (Dobkin, 2012), as shown in Fig. 3.11.

SINR at the UE used in the simulation is

$$\text{SINR}_{\text{sim}} = \frac{P_r}{N_\text{o} + \sum_{j=1}^{j=N} P_{rj}} \tag{3.68}$$

where the SINR_{sim} is the received SINR by the user calculated by the simulator; P_{rj} is the received signal from neighboring cell; and $j = \{1 \cdots N\}$, where N is the number of neighboring cells. For simplicity, first tier (six cells around the centralized cell) is used in planning for an urban area, with N_o as the background noise at the receiver.

$$\text{SINR}_{\text{sim}} = \frac{\frac{P_t G_t G_r}{L_{\text{prop}} \, L_t \, L_{\text{re}}}}{N_\text{o} + \sum_{j=1}^{j=N} \frac{P_j G_{t,j} G_r}{L_{\text{prop},j} \, L_j \, L_{\text{re}}}} \tag{3.69}$$

where L_t, L_j, and L_{re} are feeder loss for senders (central BS) and the surrounding BS$_j$ and destination. The spectral efficiency calculated by the simulation is

$$C_{sim} = 0.5\log_2(1 + SINR_{sim})$$ (3.70)

By using Eq. (3.70), the spectral efficiency is

$$C_{sim} = 0.5\log_2\left(1 + \frac{\frac{P_t G_t G_r}{L_{prop}\,L_t\,L_{re}}}{N_o + \sum_{j=1}^{j=N}\frac{P_j G_{t,j} G_r}{L_{prop,j}\,L_j\,L_{re}}}\right)$$ (3.71)

The next section explains the performance analysis of fixed and mobility system via Moving Relay (MR) to improve the throughput and capacity for vehicular users at LTE-A networks and proposing a new algorithm to minimize the power consumption for MR.

3.12 Balance Transmission Power for MR in LTE-A Cellular Networks

This section presents the performance analysis of DL and UL transmission of direct and multi-hop links. In order to enhance the RSS and throughout for vehicular users, AF relay is proposed as MR because it is related with lower delay with high speed of the vehicle than DF relay. A new algorithm to balance and minimize the transmission power consumption for MR is proposed in this section. This algorithm is based on mathematical analysis of UL and DL transmission at both direct and multi-hop links.

3.12.1 Performance Analysis of Multiusers Network

Half duplex transmission mode at the relay node proposed in this scheme is shown in Fig. 3.12. In general, the received signal at each node is based on the following relationship (Sadek et al., 2010).

$$Y = HX + n$$ (3.72)

where X is the transmitted symbol from BS, H is the coefficient channel between the source and the destination, and n is the AWGN in the corresponding channels with variance σ_o, i.e., $n \sim CN(0, \sigma_o)$. Therefore, the system performance is evaluated based on two schemes: fixed and mobility schemes.

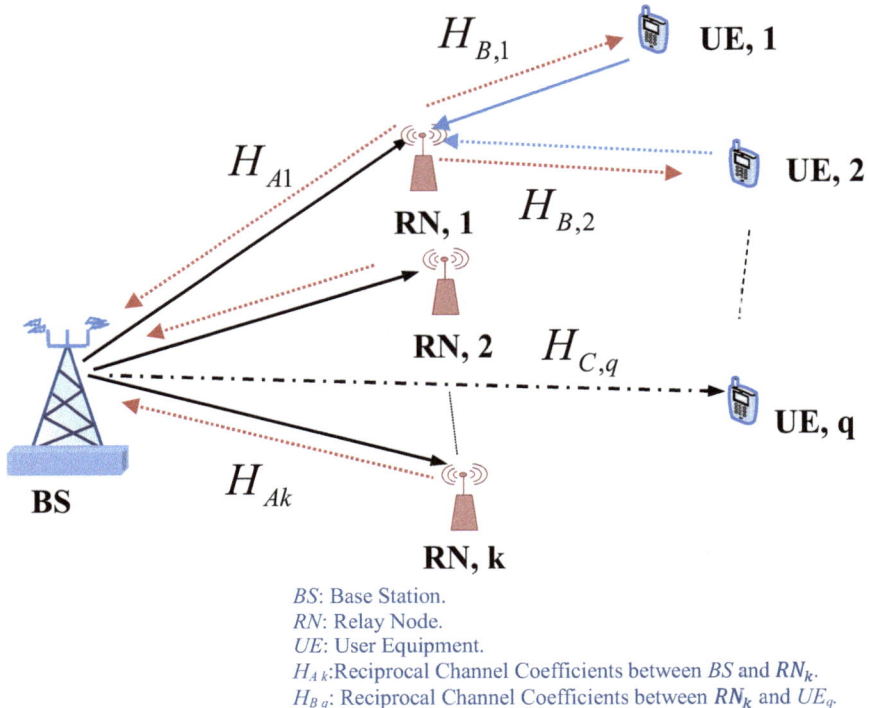

BS: Base Station.
RN: Relay Node.
UE: User Equipment.
H_{Ak}: Reciprocal Channel Coefficients between BS and RN_k.
H_{Bq}: Reciprocal Channel Coefficients between RN_k and UE_q.

Fig. 3.12 Twelve multi-hop against multiuser system at half-duplex mode

3.12.1.1   Fixed System

In this scheme, all nodes, BS, RN, and UE_q, are fixed as where the set of UE_q is $Q = \{UE_1, UE_1, \ldots UE_q\}$ where Q is set of attached users with RN and $q = 1$, 2, 3…Q. In Half Duplex (HD) mode, the relay cannot simultaneously transmit and receive. Thus, in time slot $[t_1]$, the RN receives information from both the BS and UEs, as shown in Fig. 3.12. Therefore, the received signal, $y_{RN}[t_1]$, can be written as

$$y_{RN}[t_1] = H_A X[t_1] + \sum_{q=1}^{Q} H_{B,q} X[t_1] + n_{RN} \qquad (3.73)$$

where n_{RN} is the AWGN with variance σ_o. The received signal at UE from BS via a direct link is

$$y_{UE,q}[t_1] = H_{C,q} X[t_1] + n_{UE}[t_1] \qquad (3.74)$$

At the second slot $[t_2]$, the BS and UE_q receive the amplified signals from RN with amplification factor Ψ. Therefore, to evaluate the value of this amplification factor, the in/out signals for RN must be derived, as following equations:

Firstly, for simplification let system consists of RN, UE, and BS and all noise is equally, where $n_{UE} = n_{RN} = n_{BS} = N_0$ (Sadek et al., 2010); therefore, the received signal by RN at first time slot t_1 is

$$y_{RN}[t_1] = H_A X_i[t_1] + H_B X_{UE}[t_1] + N_o \qquad (3.75)$$

Then the output signal from relay after amplification is

$$X_{RN}[t_2] = \Psi y_{RN}[t_1] + N_o \qquad (3.76)$$

By substituting Eq. (3.75) in Eq. (3.76), the result is

$$X_{RN}[t_2] = \Psi (H_A X_i[t_1] + H_B X_{UE}[t_1] + N_o) \qquad (3.77)$$

By using Expectation function on two sides (Koralov & Sinai, 2007)

$$E|X_{RN}[t_2]|^2 = E|\Psi (H_A X_i[t_1] + H_B X_{UE}[t_1] + N_o)|^2 \qquad (3.78)$$

$$P_{RN} = \Psi^2 \left[E|H_A X_i[t_1]|^2 + E|H_B X_{UE}[t_1]|^2 + E|N_o|^2 \right] \qquad (3.79)$$

$$P_{RN} = \Psi^2 \left[|H_A|^2 E|X_i[t_1]|^2 + |H_B|^2 E|X_{UE}[t_1]|^2 + N_o \right] \qquad (3.80)$$

$$P_{RN} = \Psi^2 \left[|H_A|^2 P_i + |H_B|^2 P_{UE} + E|N_o|^2 \right] \qquad (3.81)$$

$$\Psi = \sqrt{\frac{P_{RN}}{|H_A|^2 P_i + |H_B|^2 P_{UE} + N_o}} \qquad (3.82)$$

In order to study the mutual benefits for BS and UEs via relay link, the UL and DL performances are analyzed (Dahlman, Parkvall, & Skold, 2011), so the UL transmitted signal from the RN to the BS is represented as

$$y_{BS}[t_2] = \Psi H_A y_{RN}[t_1] + n_{BS} \qquad (3.83)$$

By substituting Eq. (3.74) in Eq. (3.84) (see Appendix B), the result is

$$y_{BS}[t_2] = \Psi H_A \left[H_A X[t_1] + \left(\sum_{q=1}^{Q} H_{B,q} X_q[t_1] \right) + n_{RN} \right] + n_{BS} \qquad (3.84)$$

At the UL, the BS receives two signals, via relay link and direct link, and then the BS combines these signals from multiusers and relays; therefore, Eq. (3.85) will be as

$$y_{\text{BS}}[t_2] = \left(\sum_{q=1}^{Q} H_{B,q} X_q[t_2] \right) + \Psi H_A H_A X[t_1] + \left(\Psi H_A \sum_{q=1}^{Q} H_{B,q} X_q[t_1] \right) \tag{3.85}$$
$$+ \Psi H_A n_{\text{RN}} + n_{\text{BS}}$$

while at DL the received signal at each q-UE via relay and direct link is represented as

$$y_{\text{UE},q}[t_2] = H_{C,q} X[t_2] + \Psi H_{B,q} y_{\text{RN}}[t_1] + n_{\text{UE}} \tag{3.86}$$

By substituting Eq. (3.74) in Eq. (3.87) (see Appendix C), the result is

$$y_{\text{UE},q}[t_2] = H_{C,q} X[t_2] + \Psi H_A H_{B,q} X[t_2] + \left(\Psi H_{B,q} H_{B,q} X[t_2] \right) + \Psi H_{B,q} n_{\text{RN}} + n_{\text{UE}} \tag{3.87}$$

where n_{BS} and n_{UE} are the AWGNs with variance σ_o at the BS and UE$_q$, respectively. Each source node processes and cancels the self-interface term from the received signal (Chun & Park, 2012; Khafagy et al., 2013). Therefore, the resulting signals at BS and UE$_q$ can be rewritten as

$$\widehat{y}_{\text{BS}}[t_2] = \left(\sum_{q=1}^{Q} H_{C,q} X_q[t_2] \right) + \left(\Psi H_A \sum_{q=1}^{Q} H_{B,q} X_q[t_1] \right) + \Psi H_A n_{\text{RN}} + n_{\text{BS}} \tag{3.88}$$

In addition, the signal at each q-user is expressed as

$$\widehat{y}_{\text{UE},q}[t_2] = H_{C,q} X[t_2] + \Psi H_A H_{B,q} X[t_2] + \Psi H_{B,q} n_{\text{RN}} + n_{\text{UE},q} \tag{3.89}$$

Assuming that the noise at all sources is equal $N_{\text{UE}} = N_{\text{RN}} = N_{\text{BS}} = N_0$ and based on Eqs. (3.89) and (3.90), the evaluating of instantaneous SNR in two ways (DL and UL), respectively, is as follows:

$$\rho_{\text{UE},q} = \frac{P_i |H_{C,q}|^2}{N_0} + \frac{\Psi^2 P_i |H_A|^2 |H_{B,q}|^2}{\left(\Psi^2 |H_{B,q}|^2 + 1 \right) N_0} \tag{3.90}$$

$$\rho_{\text{BS}} = \frac{\sum_{q=1}^{Q} P_{\text{UE}} |H_{C,q}|^2}{N_0} + \frac{\Psi^2 P_{\text{UE}} |H_A|^2 \sum_{q=1}^{Q} |H_{B,q}|^2}{\left(\Psi^2 |H_A|^2 + 1 \right) N_0} \tag{3.91}$$

By substituting Eq. (3.83) in Eq. (3.91) (see Appendix D), the instantaneous SNR at UE$_q$ is

$$\rho_{\text{UE},q} = \frac{P_i |H_{C,q}|^2}{N_0} + \frac{P_{\text{RN}} P_i |H_A|^2 |H_{B,q}|^2}{\left(P_{\text{RN}} |H_{B,q}|^2 + |H_A|^2 P_i + |H_B|^2 P_{\text{UE}} + N_0 \right) N_0} \tag{3.92}$$

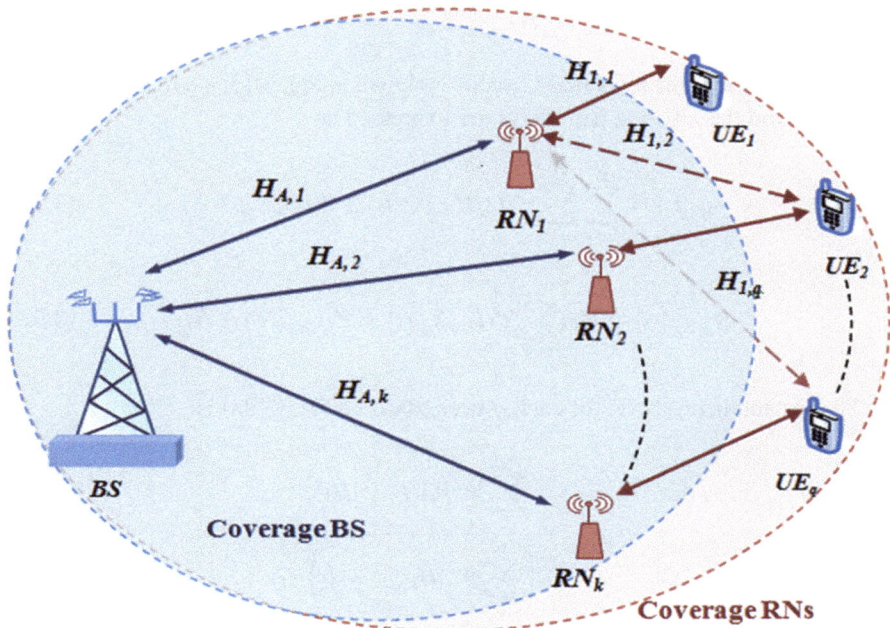

BS: Base Station.
RN: Relay Node.
UE: User Equipment.
$H_{A,k}$:Reciprocal Channel Coefficients between BS and RN_k.
$H_{k,q}$: Reciprocal Channel Coefficients between RN_k. and UE_q.

Fig. 3.13 System model of multi-RNS against multiuser system

The achievable bit rates of the multi-hop node at each qth-user is represented as

$$R_{UE,q} = \frac{1}{2}\log_2\left(1 + \rho_{UE,q}\right) \tag{3.93}$$

Similarly, for multi-RNs, Kth-relays are intermediate between Qth-users and BS as shown in Fig. 3.13. Set of RNs is $K = [RN_1, RN_2, \ldots RN_k]$, suggesting that the interference between the RNs is ignored to simplify the performance analysis. Thus, the vectors of the reciprocal channel coefficients between the nodes (RN_k, UE_q) are

$$H_{k,q} = \begin{bmatrix} H_{1,1} & H_{1,2} & \cdots & H_{1,q} \\ H_{2,1} & H_{2,2} & \cdots & H_{2,q} \\ \vdots & \vdots & \vdots & \vdots \\ H_{k,1} & H_{k,2} & \cdots & H_{k,q} \end{bmatrix} \tag{3.94}$$

while the vectors of the reciprocal channel coefficients between the nodes (BS, RN_k) are

$$H_{A,k} = \begin{bmatrix} H_{A,1} & H_{A,2} & \cdots & H_{A,k} \end{bmatrix} \tag{3.95}$$

Similarly, for multi-relay node system as shown in Fig. 3.13, the received signals at the BS and the UEs via Kth relays are expressed as

$$\hat{y}_{BS}[t_2] = \sum_{k=1}^{K} \sum_{q=1}^{Q} \Psi H_{Ak} H_{k,q} X_q[t_1] + \Psi H_{Ak} N_O + N_O \tag{3.96}$$

$$\hat{y}_{UE,q}[t_2] = H_{Ak} \sum_{k=0}^{K} \Psi H_{k,q} X_q[t_1] + \Psi H_{Bk,q} N_O + N_O \tag{3.97}$$

The instantaneous SNR for each q-user based on Eq. (3.98) is

$$\rho_{UE,q} = \frac{\sum_{k}^{K} \Psi^2 P_i |H_{Ak}|^2 |H_{k,q}|^2}{\left(\sum_{k}^{K} \Psi^2 |H_{k,q}|^2 + 1 \right) N_O} \tag{3.98}$$

By substituting Eq. (3.83) in Eq. (3.99) and performing the analysis, as well as retaining the same assumption in the previous section, the instantaneous SNR at UE_q is

$$\rho_{UE,q} = \frac{P_i P_{RN} \sum_{k}^{K} |H_{Ak}|^2 |H_{k,q}|^2}{\left(\sum_{k}^{K} P_i |H_{Ak}|^2 + 2 P_{RN} |H_{k,q}|^2 + N_O \right) N_O} \tag{3.99}$$

The achievable rates of the qth-users in this case can be represented as

$$R_{UE,q} = \frac{1}{2} \log_2 \left(1 + \rho_{UE,q} \right) \tag{3.100}$$

3.12.1.2   Mobility System

In this section, group mobility for MR and users is derived for the system performance and reducing the transmission power of MR. Derivation of the mobility group for both UEs and MRs is based on the evaluation of instantaneous SNR for both direct and relay links in Eq. (3.93).

SNR at the UE_q is the summation of SNR via BS and the SNR via MRs as follow:

$$\rho_{\text{UE},q} = \rho_{\text{UE},q}^{\text{Direct}} + \rho_{\text{UE},q}^{\text{gm}} \qquad (3.101)$$

The instantaneous SNR changes according to channel environment, such as the distance between the transmitter and receiver and fading state of the channel (Mei et al., 2013). Thus, inserting Eq. (3.5) in Eq. (3.93) produces SNR for relay link as:

$$\rho_{\text{UE},q}^{\text{gm}} = \frac{P_i P_{\text{RN}} L(d_A)^{-\alpha} |H_{k,q}|^2}{\left[P_i(d_A)^{-\alpha} + 2P_{\text{RN}} |H_{k,q}|^2 + N_O\right] N_O} \qquad (3.102)$$

and SNR via direct link is

$$\rho_{\text{UE},q}^{\text{Direct}} = \frac{P_i L(d_{C,q})^{-\alpha}}{N_O} \qquad (3.103)$$

Distance is a function of velocity and time, so the Eqs. (3.104) and (3.103) become

$$\rho_{\text{UE},q}^{\text{Direct}} = \frac{P_i L(v_{\text{UE}} T_{\text{UE}})^{-\alpha}}{N_O} \qquad (3.104)$$

$$\rho_{\text{UE},q}^{\text{gm}} = \frac{P_i P_{\text{RN}} L(v_{\text{MR}} T_{\text{MR}})^{-\alpha} |H_{k,q}|^2}{\left[P_i L(v_{\text{MR}} T_{\text{MR}})^{-\alpha} + 2P_{\text{RN}} |H_{k,q}|^2 + N_O\right] N_O} \qquad (3.105)$$

The bit rate at UE_q within group mobility is

$$R_{\text{UE},q}^{\text{gm}} = \frac{1}{2} \log_2 \left(1 + \rho_{\text{UE},q}^{\text{gm}}\right) \qquad (3.106)$$

d_A and $d_{c,q}$ are the distances between BS to each MR and UE_q, respectively, while V_{UE} and T_{UE} are the velocity and time of UE_q for direct link. V_{MR} and T_{MR} are the velocity and time of MR of relay link and $\rho_{\text{UE},q}^{\text{gm}}$ and $\rho_{\text{UE},q}^{\text{Direct}}$ are SNRs of UE within group mobility and direct link, respectively. $R_{\text{UE},q}^{\text{gm}}$ is the achievable bit rate through group mobility. Equation (3.106) describes the impact SNR with velocity variation for group mobility (MR, UE_q).

3.12.2 Balance Power Algorithm (BPA) for MR

Vehicular users suffer from outage of the wireless services especially for high speed. MR is candidate to enhance the throughput to users. However, the

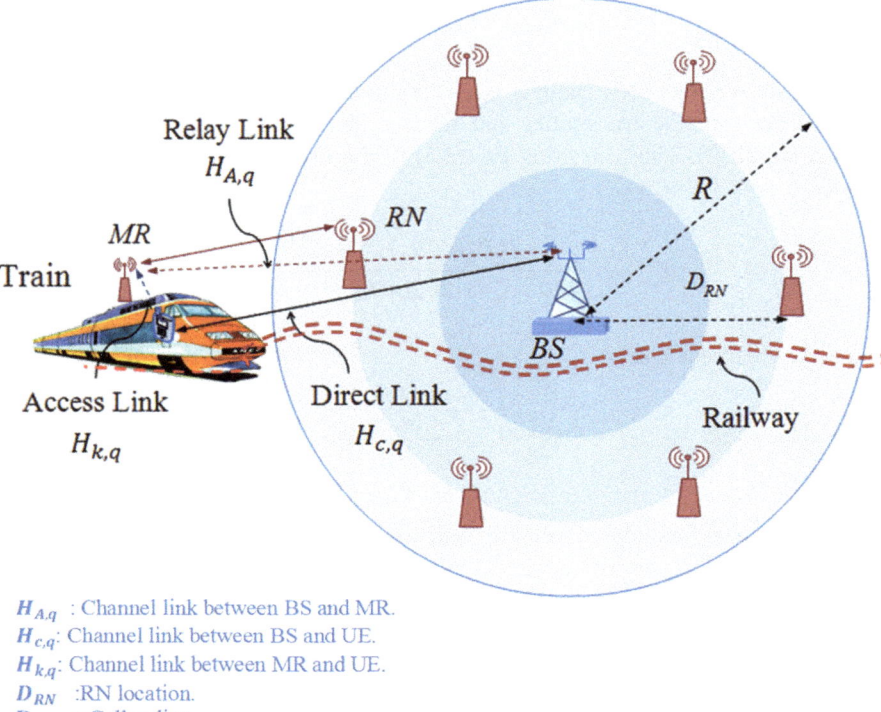

$H_{A,q}$: Channel link between BS and MR.
$H_{c,q}$: Channel link between BS and UE.
$H_{k,q}$: Channel link between MR and UE.
D_{RN} :RN location.
R   : Cell radius.

Fig. 3.14 Links of group mobility (MR fixed on the train) within travelling across the cell

power consumption reduction in wireless communications becomes an interesting issue to promote power saving with regard to environmental protection and the cost.

The proposed Balance Power Algorithm (BPA) controls transmission power for MR within cell size based on derived SNR as in Eqs. (3.103) and (3.104).

Generally, the main issue on coverage area distribution for any station is unfair distribution, where the area close to high resource links is better than at the boundaries. As a result, there is no need to use additional power to MR when the vehicle (i.e., train, bus, metro) pass near BSs and RNs as shown in Fig. 3.14.

Figure 3.14 shows the passage of the train across the cell size. Three links are considered in BPA. These links are direct link ($H_{c,q}$) between user to BS and relay link ($H_{A,q}$) between BS to MR and finally access link ($H_{k,q}$) between MR to user. $H_{c,q}$ and $H_{A,q}$ are change according to change of the distance between MR and BS.

BPA is based on the evaluation of received SNR for user during uplink and downlink transmission and provides the required SNR and throughput to user inside vehicle as well as minimizes the transmission power consumption for MR.

Algorithm 3.1: Balance Power Algorithm to Reduce and Balance

Power Consumption for MR

Require:

$V_{RN}, \rho_{th}, \rho_{max}, . P_{max}, P_{min}, Q$

Ensure:

P_{MR}

1: BEGIN

2: $Q_{MR-UE} = Q_{BS-UE} = 0$

3: **for** $i = 1$ to Q **do**

        Calculate $\rho_{UE,q}^{Direct}$, ρ_{MR}, and $\rho_{UE,q}^{Access}$

4:     **if** $\rho_{UE,q}^{Direct} \geq \rho_{UE,q}^{Access}$ **then**

5:Increment Q_{BS-UE} by 1

6:         $P_{MR} = P_{min}$

7:**else**

8:**if** $\rho_{MR} \geq \rho_{th}$ **then**

9:             Increment Q_{MR-UE} by 1

10:$P_{MR} = P_{min}$

11:**else**

12:$P_{MR} = P_{max}$

13: **for** $j = 1$ to $(\rho_{MR}/ 0.1)$ **do**

14:Calculate ρ_{MR}

15:**if** $\rho_{MR} \geq P_{max}$ **then**

16:Decrement ρ_{MR} by 0.1

17:**else**

18:Increment ρ_{MR} by 0.1

19:         **end if**

20:         **end for**

21:       **end if**

21:     **end if**

22:   **end for**

END

Inputs for this algorithm are:

ρ_{th} and ρ_{max} are the threshold and maximum required SNRs at the UE. P_{max} and P_{min} are the maximum and minimum levels of required power transmitted by the MR. Q is the number of users in the vehicle while V_{RN} is the velocity of the vehicle. Q_{MR-UE} and Q_{BS-UE} are the number of users who are attached to the MR and BS,

Fig. 3.15 Flowchart of BPA

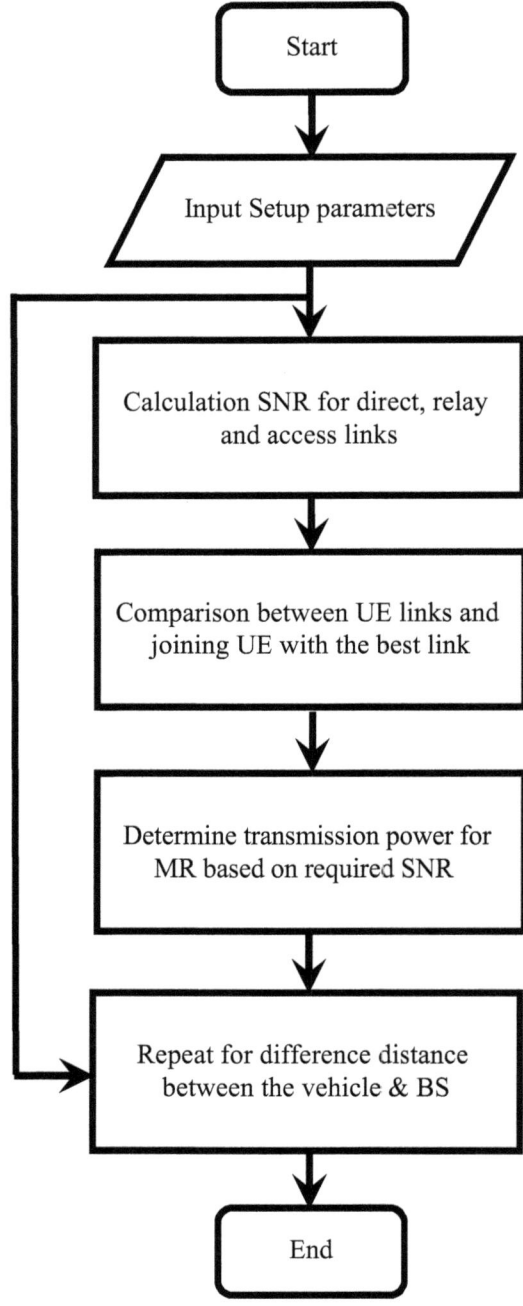

respectively. P_{MR} is the transmission power by MR. $\rho_{UE,q}^{Direct}$ and $\rho_{UE,q}^{Access}$ are the SNRs for the user at direct and access links, respectively. ρ_{MR} is the SNR for the relay link.

The main body of the balancing algorithm is described line by line in the following steps and shown in Fig. 3.15.

Line 2: select of the users attached with the MR and the users attached with the BS.

Line 3: Calculate the SNRs at the direct, relay, and access links for each user.

Line 4: Comparison between the SNRs at direct and access links and then determining the better link in order to activate it and disable the other.

Line 5: In the case that $\rho_{UE,q}^{Direct}$ is better, the user will be attached directly to the BS, so the number of users that are attached to the BS will increase by one.

Line 6: Enable the transmission power of the MR to equal the minimum chosen value (P_{min}).

Line 7-12: The algorithm proposed a second comparison between the SNR at the relay link and the required threshold SNR (ρ_{th}) of the system.

If $\rho_{MR} \geq \rho_{th}$, then the number of users that are attached to the MR will increase by one, as well as enabling the transmitted power of the MR to equal the minimum chosen value. Otherwise enable the transmitted power of the MR to equal the maximum chosen value.

Line 13-14: In this line, the counter is used in order to calculate the instantaneous value of the SNR for the MR (ρ_{MR}) for each user according to the distance between the MR and the BS.

When the MR is close to the BS, the ρ_{MR} is high, while ρ_{MR} degrades when the MR is away from the BS because of an increase in the path loss distance between the MR and the BS.

Line 15-18: Compare between the ρ_{MR} and maximum chosen SNR in order to balance and save the transmission power from the MR. This step determines the transmitted power based on the quality of the received signal for the users in the vehicle.

Line 19-22: Closed if and for statements.

3.13 Summary

This section has discussed the impact of multi-hop relay to enhance capacity and coverage area for LTE-A cellular networks. It consists of the following three main parts.

The first is called Optimum RN Deployment (ORND) and is focused on enhancing cell capacity and coverage extension at the cell edge region by the optimum deployment of RNs in the cell. In order to provide realistic capacity transmission,

The ORND is based on the mathematical formulation of modified Shannon formula. ORND consists of derivations for the saturation capacity region (X_s), the optimal location of RN (D_{RN}), the optimal number of RNs per cell (N_{relays}), and allocating the appropriate transmission power for each RN. In addition, ORND provides a solution for frequency reuse for multi-hop systems to avoid interference between RNs and the BS.

The second part details a new model called Enhance Relay Link Capacity (ERLC). This model is also based on mathematical formulation to enhance the capacity of the relay link which carries the users' information attached with the RN to the BS. ERLC employs two types of antennas: the Omni Directional (OA) antenna is used to exchange information between the RN and attached users, while the other is the directional antenna (DA) which transfers information from the RN to the BS.

Finally, the last part of this chapter demonstrated the mathematical formulation for multiuser systems for UL and DL transmission via the MR to enhance the wireless services along the travel path taken by users in public transportation vehicles. A new algorithm named BPA is proposed to minimize the consumption of transmission power for MR while travelling close to high resource links such as a BS or RN. This algorithm is represented based on a mathematical formulation of received SNR for users via the MR.

References

3GPP, TS. ETSI (2007). *Base station (BS) radio transmission and reception. 3GPP TS 36.104, Technical specification, version 8.0.0 release 8*. Valbonne, France: ETSI.

3GPP, ETSI. T. (2009). *Evolved universal terrestrial radio access (E-UTRA); Radio frequency (RF) system scenarios 3GPP ETSI 7(V8.2.0)*. Valbonne, France: ETSI.

3GPP, TS. ETSI. (2009). *Evolved universal terrestrial radio access (E-UTRA); Radio frequency (RF) system scenarios, Technical specification, version 8.2.0 release 8*. Sophia Antipolis Cedex, France: ETSI.

3GPP, TS. ETSI. (2009). *Evolved universal terrestrial radio access (E-UTRA); Physical layer procedures. Technical specification, version 8.8.0 release 8*. Sophia Antipolis Cedex, France: ETSI.

Abdulrazak, L. F., Rahman, T. A., & Sharul Kamal, A. R. (2009). *Interference reduction measurement between BWA based on MIMO over OFDM and FSS in a suburban environment*. Paper presented at the Proceedings of the 3rd IEEE International Conference on Communications and information technology, Stevens Point, WI.

Almutairi, A. F., & Salamah, S. (2006). Spectral efficiency improvement for LMDS systems using adaptive techniques in fading channels. *Computers & Electrical Engineering, 32*(4), 299–311.

Cho, S., Jang, E. W., & Cioffi, J. M. (2009). Handover in multihop cellular networks. *IEEE Communications Magazine, 47*(7), 64–73.

Chun, B., & Park, H. (2012). A spatial-domain joint-nulling method of self-interference in full-duplex relays. *IEEE Communications Letters, 16*(4), 436–438.

Dahlman, E., Parkvall, S., & Skold, J. (2011). *4G: LTE/LTE-advanced for mobile broadband: LTE/LTE-advanced for mobile broadband*. UK: Academic Press.

Dobkin, D. M. (2012). *The RF in RFID: UHF RFID in Practice*. Burlington, MA: Elsevier.

Ekiz, N., Salih, T., Kucukoner, S., & Fidanboylu, K. (2005). An overview of handoff techniques in cellular networks. *International Journal of Information Technology, 2*(3), 132–136.

Gibson, J. D. (2012). *Mobile communications handbook* (3rd ed.). Boca Raton: CRC Press.

Huang, J.-H., Wang, L.-C., Chang, C.-J., & Su, W.-S. (2010). Design of optimal relay location in two-hop cellular systems. *Wireless Networks, 16*(8), 2179–2189.

Kacerginskis, E., & Narbutaite, L. (2012). Capacity and Handover Analysis in Mobile WiMAX. *Electronics and Electrical Engineering, 119*(3), 23–28.

Khafagy, M., Ismail, A., Alouini, M.-S., & Aissa, S. (2013). On the outage performance of full-duplex selective decode-and-forward relaying. *IEEE Communications Letters, 17*(6), 1180–1183.

Khan, F. (2009). *LTE for 4G Mobile Broadband Air Interface Technologies and Performance* (1st ed.). New York: Cambridge University Press.

Kitayama, T., Hasegawa, G., Taniguchi, Y., & Nakano, H. (2013, January). *Time slot-adding algorithm for improving bottleneck link throughput in IEEE 802.16 j relay networks.* Paper presented at the 2013 International Conference on Information Networking (ICOIN), Bangkok.

Koralov, L., & Sinai, Y. G. (2007). Conditional expectations and martingales. In *Theory of Probability and Random Processes* (pp. 181–200). New York: Springer.

Korowajczuk, L. (2011). *LTE, WiMAX and WLAN Network Design, Optimization and Performance Analysis.* Reston, VA: Wiley.

Kosta, C., Hunt, B., Quddus, A., & Tafazolli, R. (2013). On interference avoidance through inter-cell interference coordination (ICIC) based on OFDMA mobile systems. *IEEE Communications Surveys & Tutorials, 15*(3), 973–995.

Laiho, J., Wacker, A., & Novosad, T. (2006). *Radio network planning and optimisation for UMTS.* New York, NY: Wiley.

Le, L., & Hossain, E. (2007). Multihop cellular networks: Potential gains, research challenges, and a resource allocation framework. *IEEE Communications Magazine, 45*(9), 66–73.

Mei, H., Bigham, J., Jiang, P., & Bodanese, E. (2013). Distributed Dynamic Frequency Allocation in Fractional Frequency Reused Relay Based Cellular Networks. *IEEE Transactions on Communications, 61*(4), 10.

Mogensen, P., Na, W., Kovács, I. Z., Frederiksen, F., Pokhariyal, A., Pedersen, K. I., et al. (2007, April). *LTE capacity compared to the shannon bound.* Paper presented at the IEEE 65th Vehicular Technology Conference, 2007. VTC2007-Spring, Dublin.

Park, K., Kang, C. G., Chang, D., Song, S., Ahn, J., & Ihm, J. (2007, August). *Relay-enhanced cellular performance of OFDMA-TDD system for mobile wireless broadband services.* Paper presented at the 16th International Conference on Computer Communications and Networks, 2007. ICCCN 2007, Honolulu, HI.

Riihonen, T., Werner, S., & Wichman, R. (2011). Mitigation of loopback self-interference in full-duplex MIMO relays. *Signal Processing, IEEE Transactions on Signal Processing, 59*(12), 5983–5993.

Sadek, A. K., Han, Z., & Liu, K. (2010). Distributed relay-assignment protocols for coverage expansion in cooperative wireless networks. *IEEE Transactions on Mobile Computing, 9*(4), 505–515.

Sesia, S., Toufik, I., & Baker, M. (2011). *LTE: The UMTS Long Term Evolution.* Chichester: Wiley.

Song, K.-B., Ekbal, A., Chung, S. T., & Cioffi, J. M. (2006). Adaptive modulation and coding (AMC) for bit-interleaved coded OFDM (BIC-OFDM). *IEEE Transactions on Wireless Communications, 5*(7), 1685–1694.

Tripathi, N. D., Reed, J. H., & VanLandinoham, H. F. (1998). Handoff in cellular systems. *IEEE, Personal Communications, 5*(6), 26–37.

Wang, F., Ghosh, A., Sankaran, C., Fleming, P., Hsieh, F., & Benes, S. (2008). Mobile WiMAX systems: Performance and evolution. *IEEE Communications Magazine, 46*(10), 41–49.

Wang, F., Ghosh, A., Sankaran, C., Fleming, P., Hsieh, F., & Benes, S. (2009). Mobile WiMAX systems: Performance and evolution. *IEEE Communications Magazine, 46*(10), 41–49.

Wang, S., Wang, J., Xu, J., Teng, Y., & Horneman, K. (2011, March). *Cooperative component carrier (re-) selection for LTE-advanced femtocells*. Paper presented at the 2011 I.E. Wireless Communications and Networking Conference (WCNC), Cancun, Quintana Roo.

Wang, H., Wang, J., & Xu, J. (2010). A method and apparatus for load balance in a relay-based multi-hop wireless network: WO Patent 2,010,115,463.

Yahya, A., Aldhaibani, J. A., Ahmed, R. B. (14/April/201). *Improvements in LTE-A cellular network, document number (PT/4602/UNIMAP/13)*.

Yang, J., Aydin, M., Zhang, J., & Maple, C. (2007). UMTS base station location planning: A mathematical model and heuristic optimisation algorithms. *IET Communications, 1*(5), 1007–1014.

Yu, Y., Murphy, S., & Murphy, L. (2008). *Planning base station and relay station locations in IEEE 802.16 j multi-hop relay networks*. Paper presented at the 5th IEEE Consumer Communications and Networking Conference, CCNC 2008, Dublin, Ireland.

Zhao, Y., Fang, X., Huang, R., & Fang, Y. (2014). Joint Interference Coordination and Load Balancing for OFDMA Multi-hop Cellular Networks. *IEEE Transactions on Mobile Computing, 13*(1), 11.

Chapter 4
Performance Enhancement of Coverage Area and Capacity for 3GPP-LTE-A Networks

4.1 Mitigating Interference Between RNs

Cell edge users face not only a high propagation loss in their own cell but also have considerable interference from neighboring cells. Therefore, most of the interferences increase when the user approaches to cell boundaries. The signal from BS degrades due to an increase in the path loss from BS, while signals as interference from neighboring cells increase. Figure 4.1 shows RSS for user along with cell diameter. A low level of RSS at cell boundaries with an increasing level of interference causes blocking for many users in this region as shown in Fig. 4.1.

The deployment of RNs around BS causes more interference in the cellular network. However, the power allocation for nodes (BS and RNs) has become one of the most effective solutions to control and mitigate interference. As a result, the improvements in spectral efficiency and RSS for each case of optimum location, as summarized in Table 4.1, are based on the following two directions in order to mitigate the interference between nodes (BS and RNs):

1. Balancing the transmitted power allocated of each RN along with N_{relays} and their locations is based on mathematical derivations in Sects. 3.6 and 3.7.
2. Proposing a new frequency reuse design by exploiting available frequencies in the different regions of RNs within cell as discussed in Sect. 3.9.

Figure 4.2 shows the mathematical and simulation results of RSS at the user for RN deployment according to the proposed scenarios summarized in Table 4.2. The improvement in RSS is nearly 32 % from cell size and from −90 to −64.2 dBm at user in cell boundaries for deploying RNs in optimum location as shown in Fig. 4.2. The level of RSS is steady around high resource links such as (BS, RN) based on the derived mathematical equations in Chap. 3.

An A-ATDI simulator which is specialized for radio planning design of wireless networks was used to validate the mathematical results. One of the options of this simulator is the use of gradient color to display of RSS distribution from stations

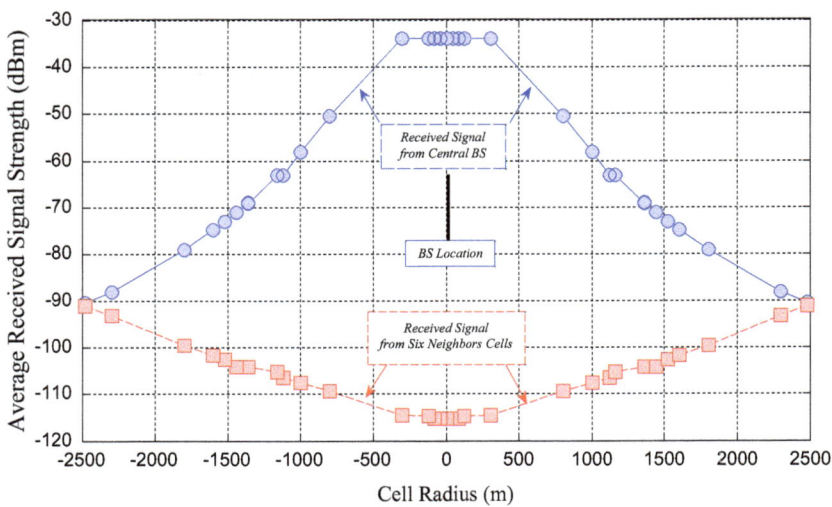

Fig. 4.1 Simulation analysis of DL RSS with interference from neighboring cells

Table 4.1 Optimum system configuration for ORND

Case	Deployment Scenario	No. of relays N_{relays}	BSTx. power P_{BS} (watt)	RNTx. power P_{RN} (watt)	RN Location D_{RN} (m)
1	4RN10W	4	40	10	1250
2	6RN5W	6	40	5	1600
3	9RN2W	9	40	2	1920

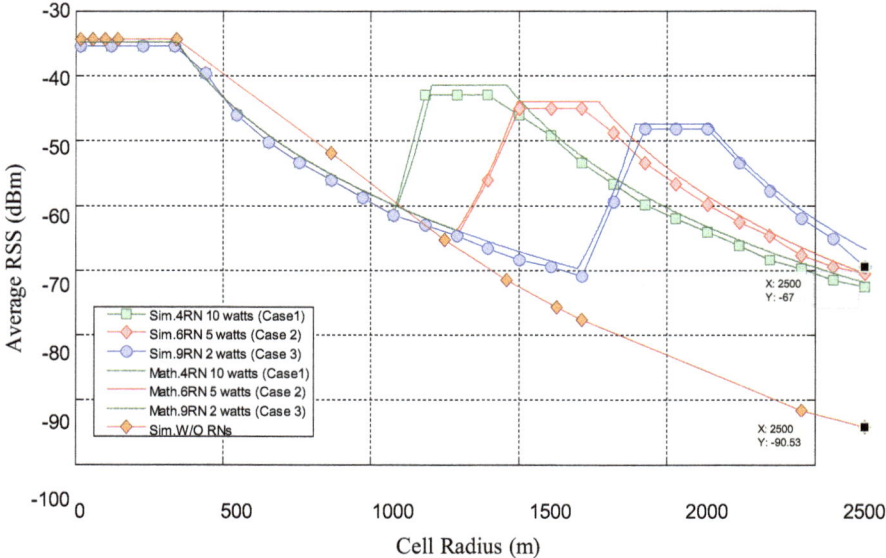

Fig. 4.2 Enhancements in RSS at UE for proposed optimum RN locations summarized in Table 4.2

Table 4.2 Simulation results of total radiation power for each optimum location

Case	Deployment scenario	BS covered area km^2	RN covered area km^2	Total covered areas km^2	Coverage area improvement (%)
1	4RN10W	9.84	4.52	14.36	31.41
2	6RN5W	9.84	4.712	14.55	32.38
3	9RN2W	9.84	4.66	14.5	32.13

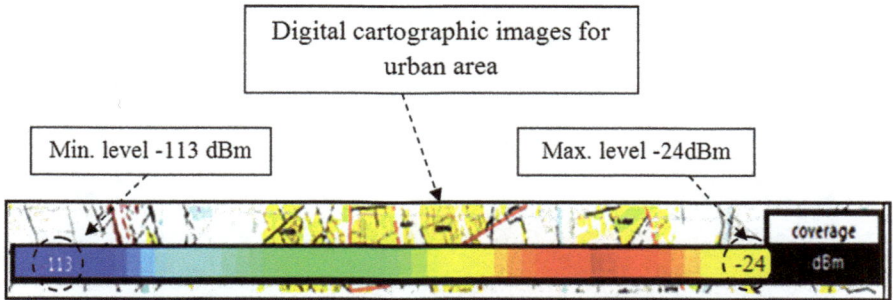

Fig. 4.3 Clarification of gradient color bar of coverage area distribution

(BSs, RNs) and shows the enhancement in the coverage area. Six traditional and identical cells (BS2, BS2 ... BS6) were deployed around central BS1 as a first-tier scheme. The frequencies were allocated according to the proposed frequency reuse scheme as mentioned in Sect. 3.9 to exploit a lower available bandwidth of frequencies. The light yellow represents maximum RSS at UE (i.e. the best coverage area), while the blue color represents the worst RSS level in gradient color schemes as shown in Fig. 4.3.

Figures 4.4, 4.5 and 4.6 describe the enhancements in coverage area distribution of the proposed locations (1250, 1600 and 1950 m, respectively). The number of RNs and their allocated powers were chosen to be compatible with each one of the locations based on Eqs. (3.53) and (3.58).

It is worth mentioning here that the most of coverage area concentrated in the area near the stations, while the poor coverage at the boundaries was caused by increasing the path loss between user and station. These figures take into consideration the interference from the first tier for cells as shown in Figs. 4.4, 4.5 and 4.6. The N_{relays} and DN$_{RN}$ and power allocated for each location in these figures were chosen according to Table 4.2. Figures 4.4, 4.5 and 4.6 show the coverage area in BS1 (center cell) changes from green to brown, which indicate the enhancements in coverage area in comparison with neighboring cells (BS2–BS7). Figure 4.7 shows the coverage area distribution for six RN where allocated power is 5 W for each RN (Case 2 in Table 4.1) in three-dimensional scheme.

Figures 4.4, 4.5 and 4.6 not only show the enhancement in coverage based on gradient color, but it is also provide validation of mathematical results in terms of

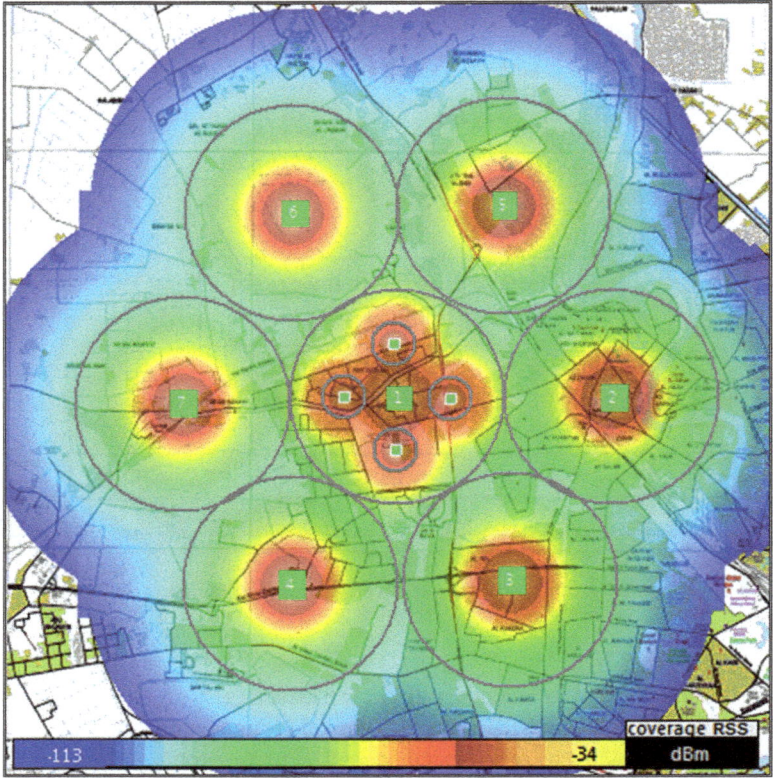

Fig. 4.4 Gradient color of coverage area distribution for four RN deployed located at 1250 m from BS with 10 W transmission power as Case 1 in Table 4.1

coverage enhancement with interference limited among RNs. In addition, the total covered areas of radiation power for proposed locations (4RN10W, 6RN5W and 9RN2W) are slightly similar in magnitude as presented in Table 4.2.

Figure 4.8 shows interference mitigation for the ORND model based on Case 2 of Table 4.1. In general, the bad interference occurs when level of received signals from the neighboring cells is equal to or exceeded the signal from a particular cell (here BS1) (Guo, Wang, & Chu, 2013; Wang et al., 2013). Figure 4.8a depicts the interference level which is represented by the pink color between each two neighboring cells before using the ORND model where the RNs in BS1 are deactivated (conventional cell). Figure 4.8b shows the dwindling interference level by using the ORND model because there is an increase in the received signal from BS1 on the interference level from neighboring cells. Therefore, the Signal-to-Interference Ratio (SIR) will increase.

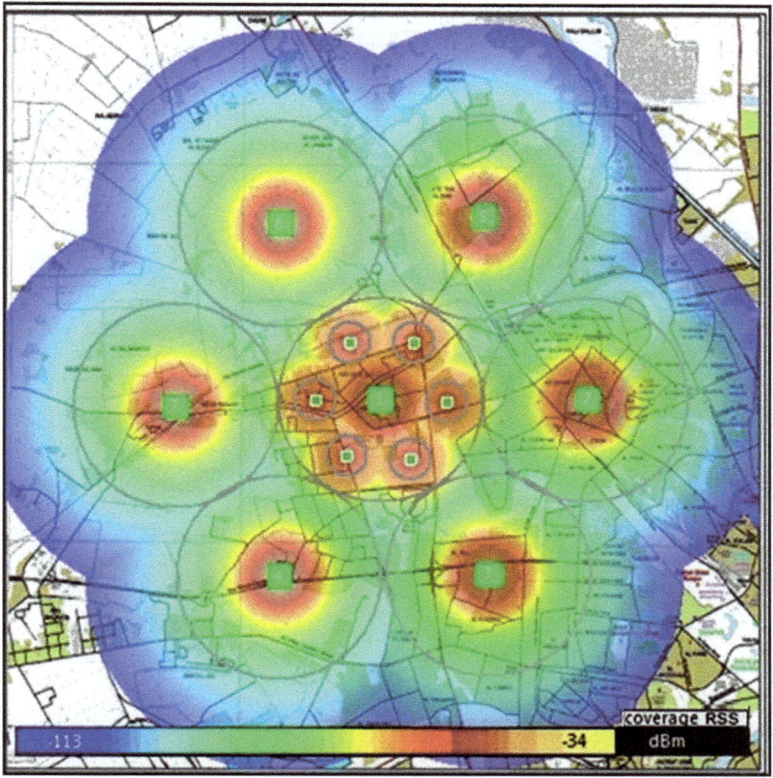

Fig. 4.5 Gradient color of coverage area distribution for six RN deployed and located at 1600 m from BS with 5 W transmission power as Case 2 in Table 4.1 (Aldhaibani, Ahmad, Yahya, Azeez, & Abbas, 2014; Aldhaibani, Yahya, Ahmad, & Md Zain, 2013)

4.2   Relay Link Enhancement

To solve the problem of capacity limited for relay link, ERLC is proposed to accommodate the number of simultaneous users generated from RN deployment at cell edge regions as discussed in details in Sect. 3.10. ERLC model employs two antennas at RN: one is the directional antenna (DA) type directed the toward the BS in order to improve relay link capacity and to achieve maximum relay location range, while the second is OA to exchange information between the RN and attached users. The following subsection analyses ERLC based on handover process evaluation.

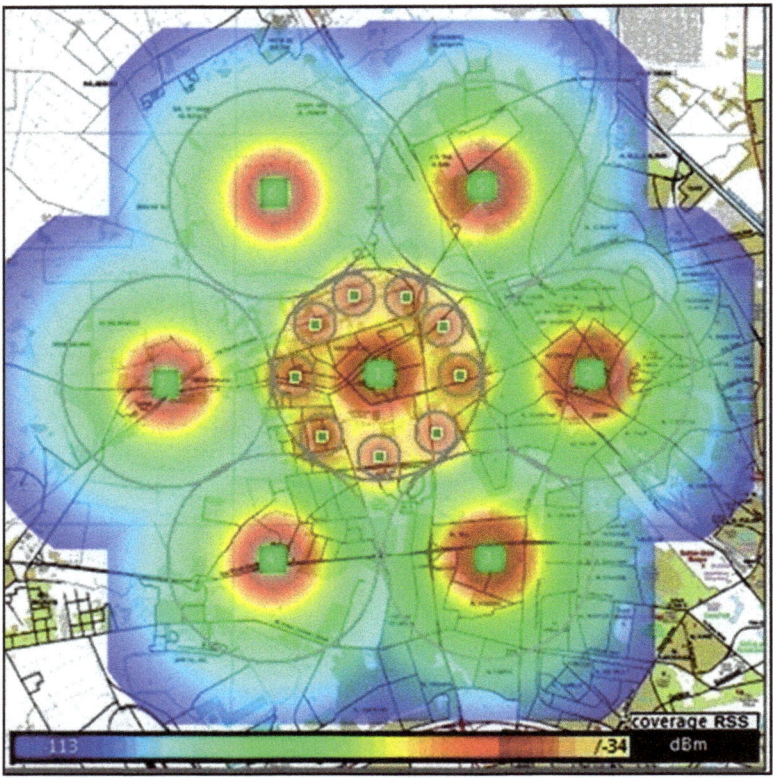

Fig. 4.6 Gradient color of coverage area distribution for nine RN deployed located at 1950 m from BS with 2 W transmission power as Case 3 in Table 4.1 (Aldhaibani, Yahya, Ahmad, & Md Zain 2013; Aldhaibani, Yahya, Ahmad, Fayadh, & Abbas, 2014)

4.2.1   Performance Analysis of Handover Process

Figure 4.9 shows the mathematical analysis of handover performance with various antenna gain of DA for RN using DA directed toward BS for different cases of α based on Eq. (3.62). Figure 4.9 illustrates that the handover location decreases toward higher resource links (i.e. BS) with on increasing DA gain value. In other words, the handover location is 1057 m with 5 dBi DA gain. However, the distance approaching BS to be 739.3 m with 10 dBi DA gain as shown in Figure 4.9. It should be noted here that any approximation in handover distance toward awarding high capacity links leads to an increase in the capacity of the relay link (Ni, Collings, & Liu, 2012).

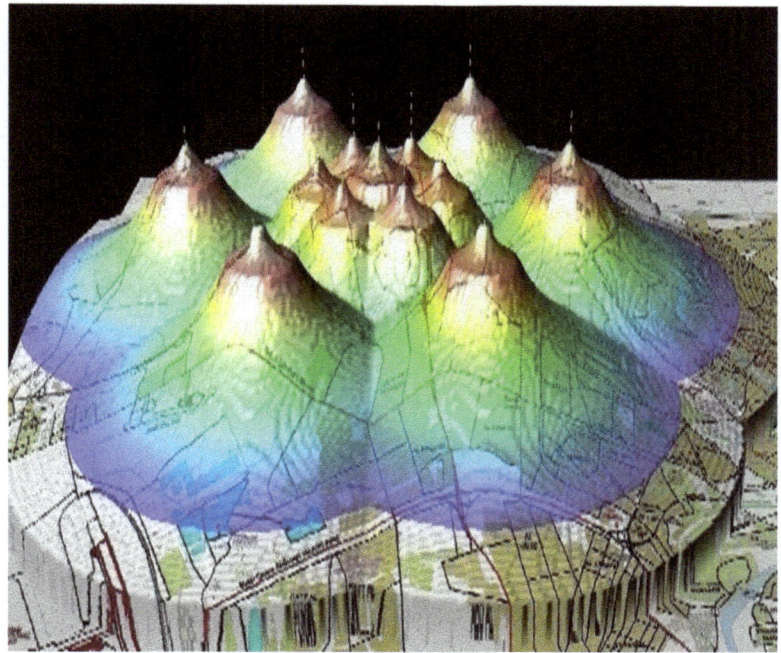

Fig. 4.7 Three-dimension gradient color of coverage area distribution for six RN deployed at proposed location with 6 W transmission power allocated Case 2 in Table 4.1 (Aldhaibani, Yahya, Ahmed, Omar, & Ali, 2013; Aldhaibani, Yahya, Ahmed, Ali, & Fayadh, 2014

4.2.2 Performance Enhancement for Relay Link

The mathematical and simulation results of the improvement of spectral efficiency and throughput at relay link are presented in this section as discussed in Sect. 3.10. Based on the proposed optimum location as discussed in Sect. 3.6, scenario 6RN5W was chosen from Table 4.1 (Scenario 2) as one of optimal locations of RN to evaluate the ERLC model.

Three different scenarios of DA gain and RN transmitted power are proposed to evaluate the enhancement in the relay link. Figure 4.10 shows three scenarios of enhancements in spectral efficiency based on Eq. (3.64) as discussed in Sect. 3.10.2. These scenarios and enhancements of levels are summarized in Table 4.3. Although there were improvements through the adoption of Scenario 1 and Scenario 2 in spectral efficiency, there was also extra power consumption in DA. Scenario 3 demonstrates the best solution for improving the relay link because the feeder power is divided between two antennas, DA and OA. As a result, Scenario 3 avoids extra consumption in the total feeder power for MR as well as achieves an enhancement in relay link as shown in Fig. 4.10 and Table 4.3.

The enhancement of spectral efficiency at the relay link according to the handover point and based on Eqs. (3.62) and (3.64) is from 1.99 to 2.95 bps/Hz

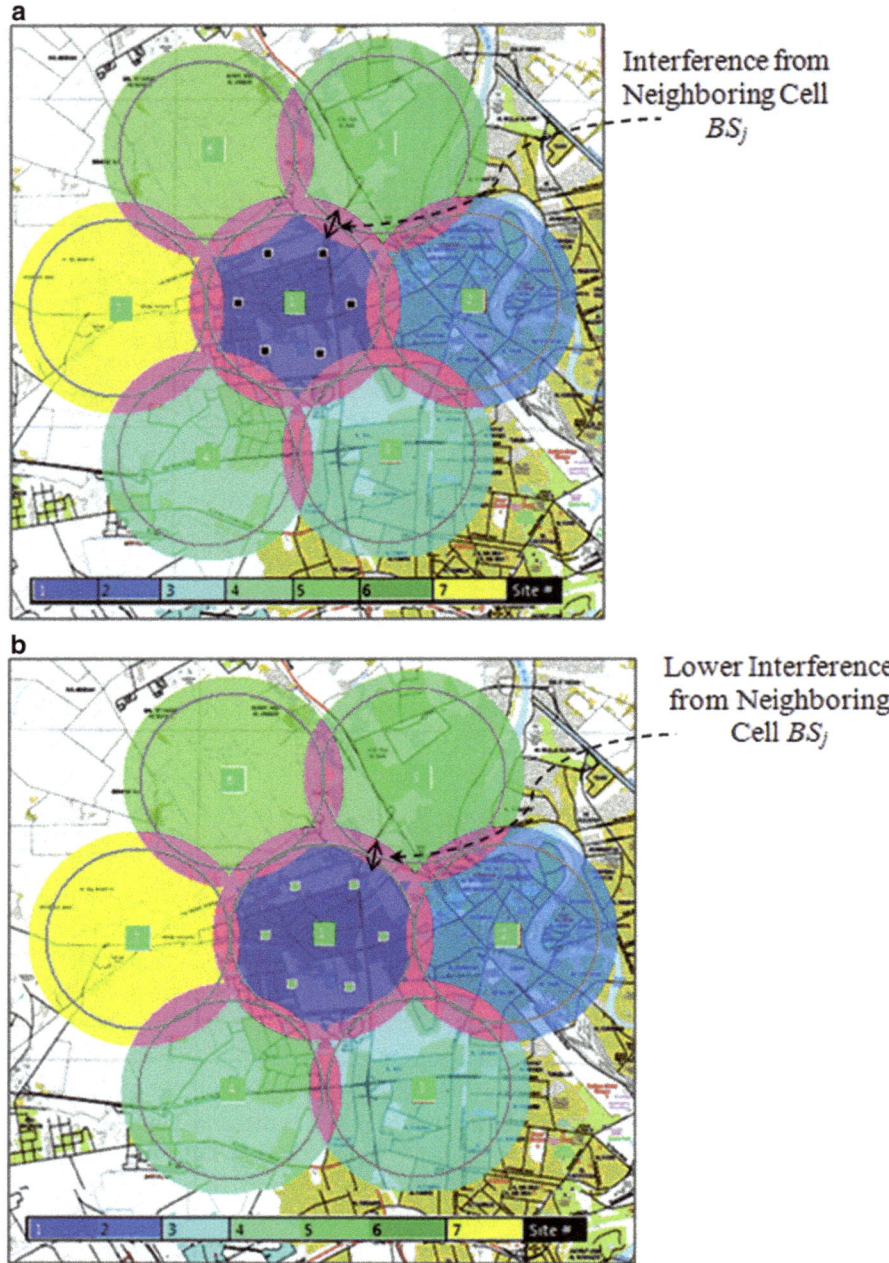

Fig. 4.8 The interference mitigation for ORND using six RN deployed with 5 W transmission power Case 2 Table 4.1 (**a**) without ORND, RNs are de-active (**b**) for ORND the RNs are active

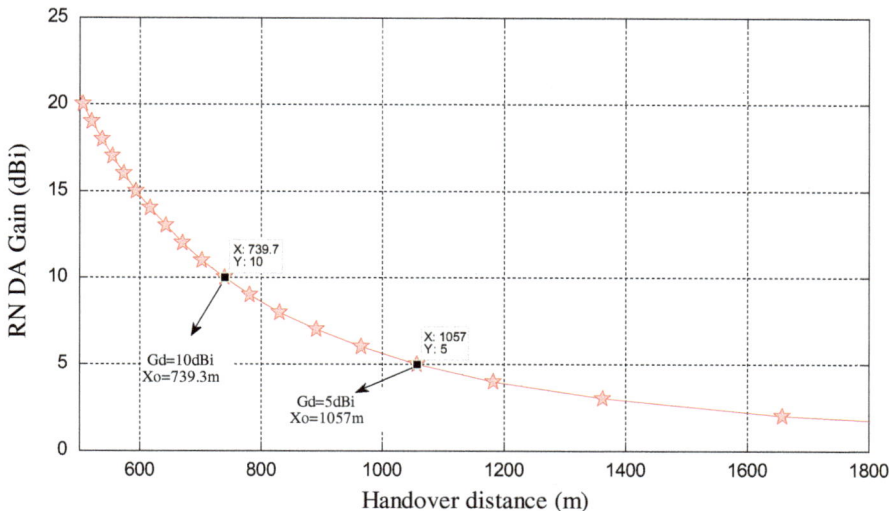

Fig. 4.9 The changes of handover distance with the different level directional antenna gain

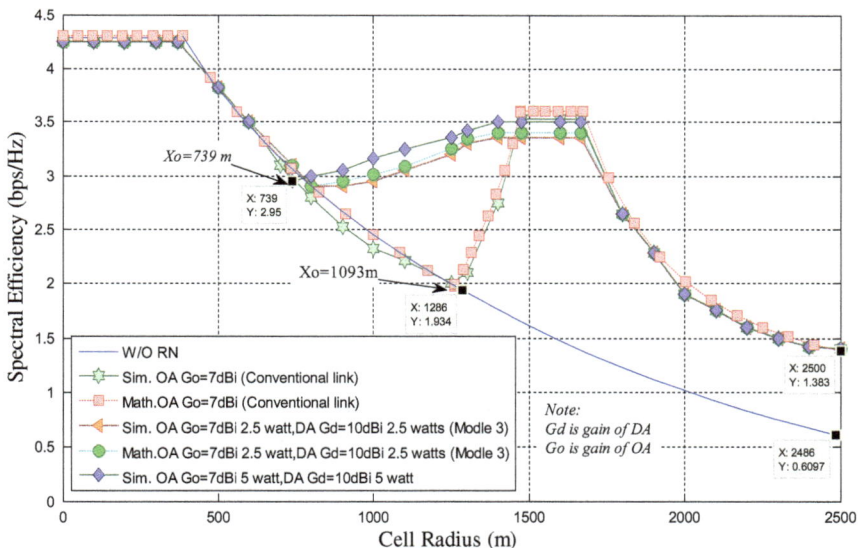

Fig. 4.10 Relay link enhancement by using DA at RN for ERLC proposed and conventional links

compared with a conventional link that used one antenna is OA for RN (i.e. 48 %
enhancement in relay link). This improvement enables the relay link to accommodate a number of active users at the cell boundaries. As a result, ERLC increases the
overall cell spectral efficiency to approximately 32 % as shown in Fig. 4.10.

Figure 4.11 shows the enhancement in throughput at UEs at relay link with
1.4 MHz bandwidth according to Eq. (3.64). Throughput at UE increased from 2.24

Table 4.3 Proposed parameters for ERLC

Scenarios	Power (watt)		Gain (dBi)		Enhancement in capacity (%)
	OA	DA	OA	DA	
Scenario 1	5	5	7	10	37
Scenario 2	5	5	7	5	29.3
Scenario 3	2.5	2.5	7	10	32

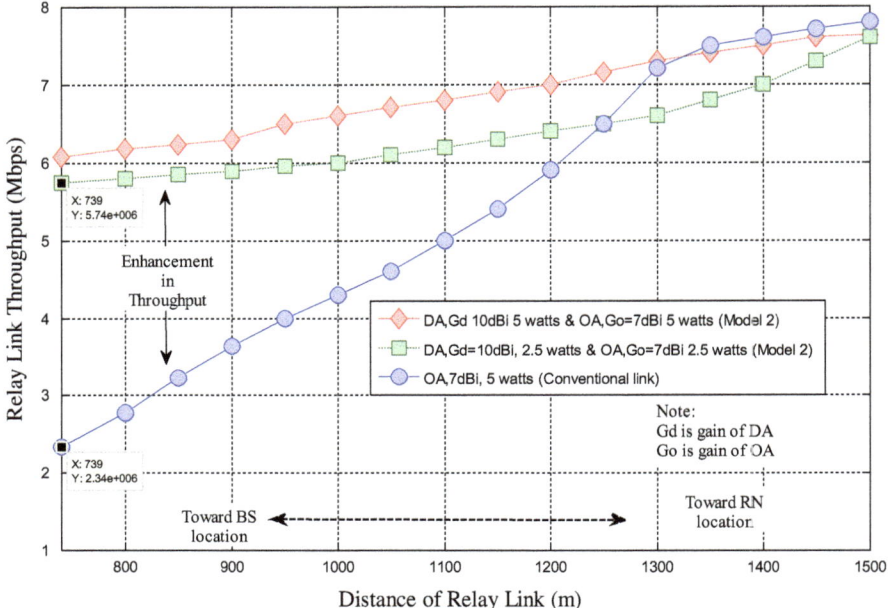

Fig. 4.11 Throughput enhancement at relay link

to 5.7 Mbps at relay link when the handover was located in 739 m from BS by using ERLC model as discussed in Sect. 3.10 and using parameters summarized in Table 4.2 Scenario 3. Throughput at UE increased whenever it approached the RN as shown in Fig. 4.11. ERLC not only increases the capacity and throughput of relay link but also reduces the outage probability and enables the relay link to overcome the environmental fluctuations for the channel.

Simulation analysis verified the mathematical results using an A-ATDI simulator which handles several propagation loss parameters based on Eq. (3.67). Figure 4.12 shows the gradient color of RSS distribution for BS with six RNs deployed around BS, so that each RN used OA for RN (i.e. conventional link). The figure shows the low level of RSS at relay link regions, which are illustrated in middle zones between BS and RNs. The low level of RSS at relay links in Fig. 4.12 causes an inability to absorb all information generated by users and sent it to BS and increases outage probability at the relay link.

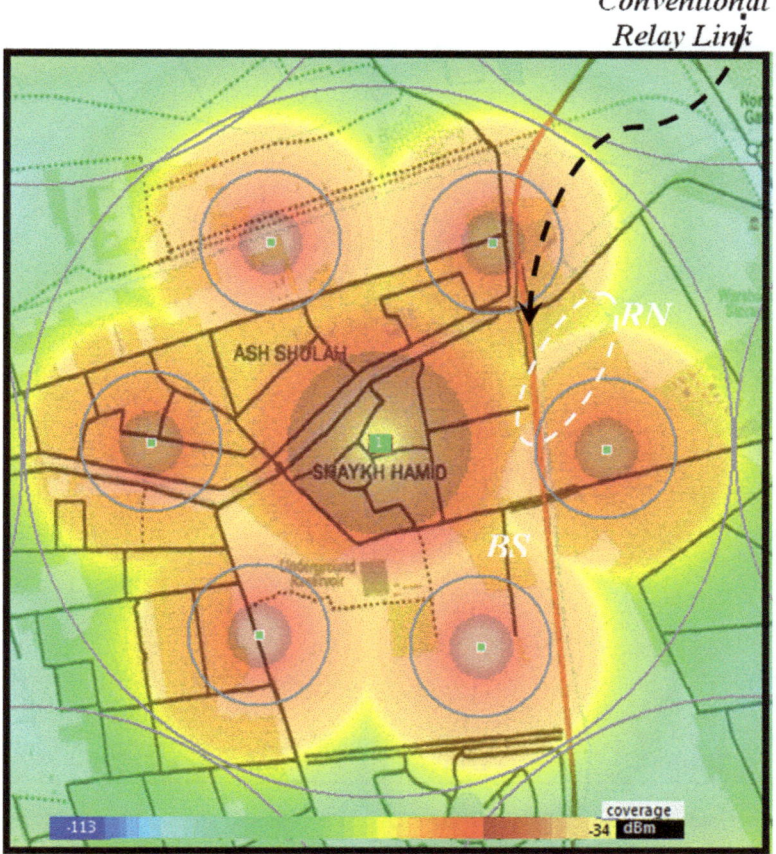

Fig. 4.12 Gradient color of RSS distribution for six RN deployed around BS each RN uses OA 7 dBi gain and 5 W feeder power (conventional link)

Figure 4.13 shows the gradient color of RSS distribution from one BS with six RNs. Each RN uses two antenna designs: one OA with 7 dBi antenna gain and the second DA designed with a 10 dBi antenna gain. The total power delivered is 5 W divided between them is summarized in Table 4.2 Scenario 3. DA was choses as broadband according to (Dobkin, 2012) as discussed in Sect. 3.10.3.

Transmission power for each RN was equally divided between OA and DA to avoid extra power consumption, based on Scenario 2 of Table 4.3. Figure 4.13 shows an improvement in RSS at the relay link for the proposed ERLC model in comparison with the conventional relay link as shown in Fig. 4.12 which uses one OA antenna.

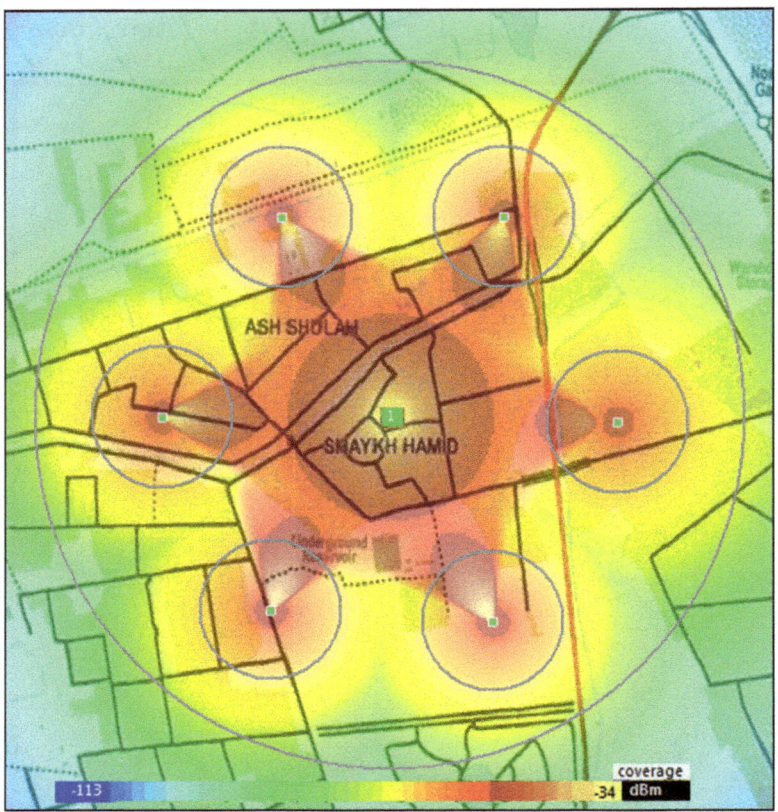

Fig. 4.13 Gradient color of RSS distribution for six RN deployed around BS each RN uses two antenna designs according to Table 4.3, Scenario 3

4.2.3   Performance Analysis of Balance Transmission Power for MR

MR is a moving relay node installed on the top of a vehicle to enhance the throughput in wireless services for the passengers when the vehicle passes within areas which have a low level of RSS, especially at cell boundaries as discussed in Sect. 2.4.

To reduce the power consumption for MR when the vehicle moves close to high resources links (i.e. BS and RN), a new algorithm is proposed. The algorithm is called the Balance Power Algorithm (BPA) and is based on performance analysis of UL and DL transmission mode for UEs inside the vehicle as discussed in details in Sect. 3.12.

The following section discusses the UL and DL transmission results for fixed and mobility scenarios and the reduction in transmission power consumption for MR based on the mathematical derivations in Sect. 3.12.

4.3   UL and DL Performance Analysis

Typically, UL and DL data transmissions are not symmetrical, especially with HD mode because of differences in technical specifications such as transceiver design and antenna for both BS and UEs. Therefore, in order to obtain mutual benefits for both BS and UEs via relay, the performance of the relay system should be studied in two ways: UL and DL transmission modes. Propagation loss effects the level of power transmitted from BS; therefore, there are low levels of RSS at the cell edge region.

Figure 4.14 shows the RSS at DL and UL mode between the UE and BS without using RNs. Propagation loss decreases the RSS level at UE of both UL and DL, while increasing the path loss between UE and BS. Most users suffer from poor wireless services at the cell boundaries as shown in Fig. 4.14.

Employing intermediate relay nodes reduces the path loss effect by dividing the propagation path between the BS and user in to two parts and provide transmit power gains (Pabst et al., 2004). Figure 4.15 shows the numerical analysis of the improvement in bit rate by using RN rather than direct transmission based on Eqs. (3.107). The improvement which occurs is in SNR and the spectral efficiency increases as the number of deployed RNs increases. That is due to the increase in resource grants in the network (Qian Li, Qian, & Geng, 2013). In Fig. 4.15, for example, at 20 dB SNR the spectral efficiency increased from 1.06 bps/Hz in direct link to 15.52 bps/Hz at multihop relay with six relays deployed ($K = 6$) as shown in Fig. 4.15. However, increasing the number of deployed RNs leads to a greater amount of resource allocation, interference and cost effect.

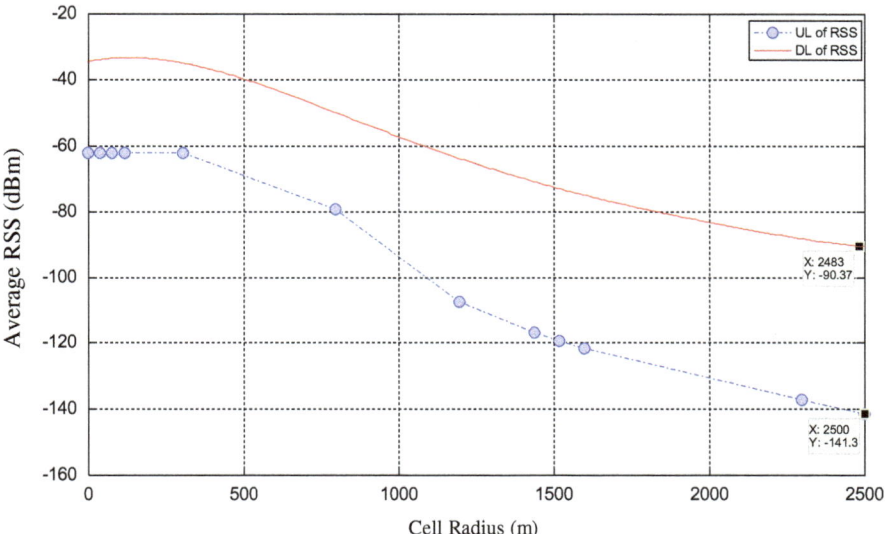

Fig. 4.14 RSS at UE vs. the cell radius within UL and DL mode for conventional cellular

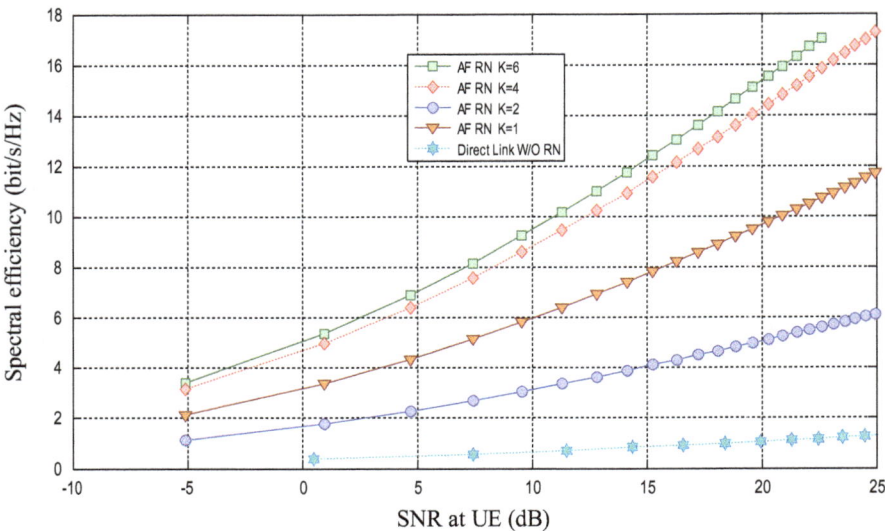

Fig. 4.15 Spectral efficiency and SNR enhancements using different number of relays

4.3.1   Performance Analysis of BPA at MR

In order to explain the performance of BPA, the numerical results of BPA at MR can be divided into two parts: the first when using BPA at MR within a conventional cellular network (i.e. the cellular that does not contain RN), while the second part is within LTE-A cellular networks when deploying six RN around the BS based on Case 2 in Table 4.2.

Figure 4.16 illustrates the RSS enhancement of UE along the cell diameter for six RNs deployed at 1600 m from BS according to Table 4.1 Case 2. The enhancement in RSS level is from −90.7 to −69 dBm at cell boundaries after RN deployment. This enhancement in RSS level avoids disconnection in wireless services and increases the number of accepted users at cell edge region within the network.

4.4   Summary

Multihop relay is an important issue which has been introduced by LTE-A networks to enhance both coverage area and capacity for 3GPP-LTE networks. RN is considered to be an appropriate solution to address low SINR at the cell edge and to meet the access requirement of nonuniform distributed traffic in densely populated areas to improve coverage and capacity. These improvements are based on RN location.

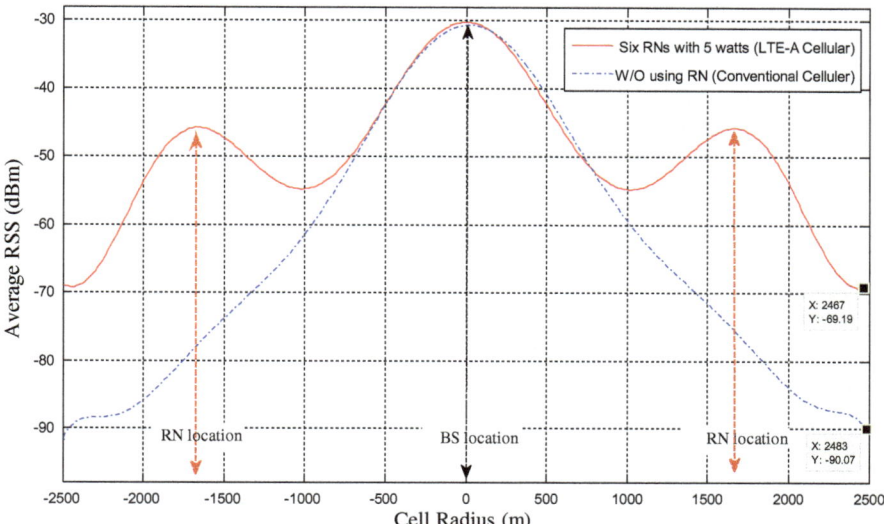

Fig. 4.16 RSS at UE vs the cell size for conventional and LTE-A contained of six RN deployed at 1600 m from BS of each RN with 5 W transmission power

ERLC model enhances the relay link capacity in order to increase the number of simultaneous users at the cell edge. This model uses two antenna designs: the DA type directed toward the BS to improve relay link quality, while the second is OA to exchange information between the RN and attached users. The results showed that there was an improvement in spectral efficiency of up 48 % in comparison with conventional relay link that uses one OA antenna.

Balance Power Algorithm (BPA) aims to balance the transmission power from MR when the vehicle moves close to high resource links such as BSs or RNs. This algorithm is based on the performance analysis for UEs inside the vehicle at UL and DL transmission.

References

Aldhaibani, J. A., Ahmad, R. B., Yahya, A., Azeez, S. A., & Abbas, A. H. (2014). Performance analysis of amplify and forward relay during uplink and downlink in LTE-A cellular networks. *Journal of Next Generation Information Technology, 5*(1), 1–8.

Aldhaibani, J. A., Yahya, A., Ahmad, R. B., Fayadh, R. A., & Abbas, A. H. (2014). Reducing transmitted power of moving relay node in LTE-A cellular networks. *Journal of Computer Science, 10*(6), 1051–1061.

Aldhaibani, J. A., Yahya, A., Ahmad, R. B., Md Zain, A. S., & Salman, M. K. (2013). On coverage analysis for LTE-A cellular networks. *International Journal of Engineering and Technology, 5*(1), 492–497.

Aldhaibani, J. A., Yahya, A., Ahmad, R. B., & Md Zain, A. S. (2013). performance analysis of two-way multi-user with balance transmitted power of relay in LTE-A cellular networks. *Journal of Theoretical and Applied Information Technology, 51*(1), 183–190.

Aldhaibani, J. M., Yahya, A., Ahmed, R. B., Ali, Z. G., & Fayadh, R. A. (2014). Enhancing link quality in a multi-hop relay in LTE-A employing directional antenna. In *2013 I.E. International RF and Microwave Conference*, Penang, 9–11 December 2013. New York: IEEE.

Aldhaibani, J. A., Yahya, A., Ahmed, R. B., Omar, N., & Ali, Z. G. (2013). Effect of relay location on two-way DF and AF relay for milt-users system in LTE-A cellular networks. In *Business Engineering and Industrial Applications Colloquium (BEIAC), 2013 I.E. International Conference*, 7–9 April 2013, Langkawi, Malaysia. New York: IEEE.

Aldhaibani, J. A., Yahya, A., Ahmed, R. B., & Azeez, S. A. (2014). Increasing the coverage area through relay node deployment in long term evolution advanced (LTE-A) cellular networks. In M. F. Ramli, A. K. Junoh, N. Roslan, M. J. Masnan, & M. H. Kharuddin (Eds.), *International Conference on Mathematics, Engineering & Industrial Applications 2014*, Penang, Malaysia, 28–30 May 2014. Melville: American Institute of Physics (AIP).

Aldhaibani, J. A., Yahya, A., & Ahmad, R. B. (2014). Coverage extension and balancing the transmitted power of the moving relay node at LTE-A cellular network. *The Scientific World Journal, 2014*(815720), 1–10.

Dobkin, D. M. (2012). *The RF in RFID: UHF RFID in practice*. Burlington, MA: Elsevier.

Guo, W., Wang, S., & Chu, X. (2013). *Capacity expression and power allocation for arbitrary modulation and coding rates*. Paper presented at the IEEE Wireless Communications and Networking, Shanghai.

Ni, W., Collings, I. B., & Liu, R. P. (2012). Relay handover and link adaptation design for fixed relays in IMT-advanced using a new Markov chain model. *IEEE Transactions on Vehicular Technology, 61*(4), 1839–1853.

Pabst, R., Walke, B. H., Schultz, D. C., Herhold, P., Yanikomeroglu, H., Mukherjee, S., et al. (2004). Relay-based deployment concepts for wireless and mobile broadband radio. *IEEE Communications Magazine, 42*(9), 80–89.

Qian Li, R. Q. H., Qian, Y., & Geng, W. (2013). Intracell cooperation and resource allocation in a heterogeneous network with relays. *IEEE Transactions on Vehicular Technology, 62*(4), 1770–1784.

Wang, J., Huang, Y., Zhong, C., Al-Qahtani, F., Wu, Q., & Cheng, Y. (2013). Performance analysis of interference-limited dual-hop multiple antenna AF relaying systems with feedback delay. *EURASIP Journal on Wireless Communications and Networking, 2013*(1), 1–15.

Chapter 5
Optimum Location for Relay Node in LTE-A

5.1 Conclusion

Multi-hop relay is considered to be one of the main aspects of the LTE-A to meet the growing demand for coverage extension and capacity enhancement by improving LTE performance. It has recently been the center of attention by the wireless communications community.

This chapter listed the contributions of this work and provides suggestions for possible future works. Four main topics related to multi-hop relay in LTE-A cellular networks are covered.

A new model to determine the optimum location for RN, the (D_{RN}), in LTE-A cellular networks is formulated. In order to provide flexibility to radio planning design, three optimum locations (1250, 1600, and 1950 m) are derived based on the proposed mathematical modeling analysis. These locations have shown a 40 % improvement in capacity at cell edge regions in comparison with a conventional cell (i.e., the cell that does not contain RN). As a result, the gain in capacity is almost the same for any chosen location from them. The interference mitigation between stations (BSs and RNs) is considered using mathematical analysis. The modified Shannon formula for LTE downlink capacity distribution is considered in order to provide a realistic transmission in comparison with classical Shannon formula that has been considered by most previous researches (Chen, 2012; Guo, Wang, & Chu, 2013; Mogensen et al., 2007) as presented in Appendix A.

The optimum number of relays (N_{relays}) is derived based on RN transmission power and path loss analysis among stations (BSs and RNs) to alleviate the overlapping among neighboring RNs while providing best coverage. The transmission power for each RN location is allocated based on mathematical formulations to avoid the overlapping between RNs and to ensure lower power consumption with capacity enhancement at the cell edge region. (N_{relays}) is almost totally dependent on the (D_{RN}) and transmission power of both RNs and the BS, so that a higher RN transmission power means fewer RNs should be deployed in the cell to minimize

© Springer International Publishing Switzerland 2017

A. Yahya, *LTE-A Cellular Networks*, DOI 10.1007/978-3-319-43304-2_5

mutual overlapping. A new method of frequency reuse for multi-hop relay is designed to suppress the interference among stations (RNs and BS) by exploiting the same available spectrum for the cell.

The quality of relay link is improved by using a new model which employs two types of antenna for the RN. Based on the analysis, the capacity of relay link is increased by 46% in comparison with conventional design consisting of one OA. This enables the relay link to accommodate the increased number of users at the cell edge region. The transmitted power for each antenna is allocated so as to provide the best enhancement for the relay link by maintaining the same power consumption of the feeder power for the RN.

The MR is proposed and evaluated to improve RSS and throughput for passengers on public transportation. In addition, the consumption of transmission power for the MR was substantially reduced by proposing the BPA. In order to achieve maximum yields, the BPA is based on an evaluation of UL and DL transmission mode for mobility multi-hop systems within the LTE-A cell (i.e., the cell containing the deployed RN). The BPA creates a balance between transmission power consumption for MR and required throughput for the users. The results showed that by using BPA, the consumption of transmission power is reduced by 75 % and an increase in the number of simultaneous users attached with the MR in comparison with direct links without the MR.

This book derived and provided formulas for radio planning designers to achieve best coverage and capacity by using RNs for LTE cellular networks. These formulas will assist designers' indetermination, the (D_{RN}), (N_{relays}), and resource allocations for each RN without using licensed and expensive simulators.

References

Chen, G. (2012). *Rate enhancement and multi-relay selection schemes for application in wireless cooperative networks*. Loughborough: Loughborough University.

Guo, W., Wang, S., & Chu, X. (2013, April). *Capacity expression and power allocation for arbitrary modulation and coding rates*. Paper presented at the IEEE Wireless Communications and Networking, Shanghai, China.

Mogensen, P., Na, W., Kovács, I. Z., Frederiksen, F., Pokhariyal, A., Pedersen, K. I., et al. (2007, April). *LTE capacity compared to the shannon bound*. Paper presented at the IEEE 65th Vehicular Technology Conference, 2007. VTC2007-Spring, Dublin.

Appendix-A

Modification for Shannon Formula

In order to facilitate accurate benchmarking of Universal Terrestrial Radio Access Network (UTRAN) Long Term Evolution (LTE), the Shannon capacity bound is modified. The modification in Shannon capacity bound is generally applied to wireless communication systems, not only to LTE system.

Recall the classical SISO Shannon capacity formula for the theoretical channel spectral efficiency as a function of SINR:

$$C = \log_2(1 + \text{SINR}) \qquad (A.1)$$

where C is the capacity in classical Shannon formula in (bps/Hz). This formula is valid for infinite delay and infinite code block size in an AWGN channel. For general MIMO with perfect transmitted knowledge, the Shannon capacity is (Wang et al., 2008, 2011; Wang, 2010).

$$C = \sum_{ka=1}^{\min(n_T, n_R)} \log_2(1 + \text{SINR}_{ka}) \qquad (A.2)$$

Here n_T and n_R denote the number of transmit and receive antennas, respectively, and SINR_{ka} denotes the resulting SNR of the ka_{th} spatial sub-channel which is influenced by the Eigen value, the noise/interference, as well as the allocated transmit power on that sub-channel. The Shannon Capacity bound in Eq. (D.1) cannot be reached in practice due to several implementation issues.

To evaluate LTE link level, two parameters are chosen to fitting performance of Shannon capacity bound: Bandwidth efficiency and the SINR efficiency (Wang et al., 2009, 2011; Wang, 2010).

Table A.1 CQI table for modulation and coding schemes in LTE networks (3GPP, 2009; Sesia et al., 2011)

CQI	Modulation scheme	Coding rate	Spectral efficiency (bps/Hz)	SINR (dB)
0	QPSK	0.076	0.1523	−2.5
1	QPSK	0.12	0.2344	−0.4
2	QPSK	0.19	0.3770	0.8
3	QPSK	0.3	0.6016	1.5
4	QPSK	0.44	0.877	4.5
5	QPSK	0.59	1.1758	6.8
6	16QAM	0.37	1.4766	9
7	16QAM	0.48	1.914	10.7
8	16QAM	0.6	2.406	13
9	16QAM	0.45	2.7305	13.8
10	64QAM	0.55	3.223	14.3
11	64QAM	0.65	3.903	15
12	64QAM	0.75	4.312	18
13	64QAM	0.65	3.75	19
14	64QAM	0.85	5.11	21.5
15	64QAM	0.93	5.554	22.6

These parameters used to modify Shannon capacity formula, therefore the modified Shannon formula is

$$C = \mathrm{BW}_{\mathrm{eff}} \log_2(1 + \rho_i/\rho_{\mathrm{eff}}) \qquad (A.3)$$

Here $\mathrm{BW}_{\mathrm{eff}}$ adjusts for the system bandwidth efficiency of LTE and ρ_{eff} adjusts for the SINR implementation efficiency of LTE. The upper limit C is chosen according to the hard spectral efficiency given by Modulation and Coding Scheme (MCS), e.g. 64QAM, Rate 4/5 for the single stream case.

The list of MCS with CQI values supported by 3GPP LTE standards (3GPP, 2009; Sesia et al., 2011) is presented in Table A.1.

A given MCS requires a certain SNIR to operate with an acceptably low Bit Error Rate (BER) in the output data. An MCS with a higher throughput needs a higher SNIR to operate. A code set contains many MCSs and is designed to cover a range of SINRs.

Figures A.1 and A.2 show the differences in performances between the conventional and modified Shannon capacity bound (Fig. A.3).

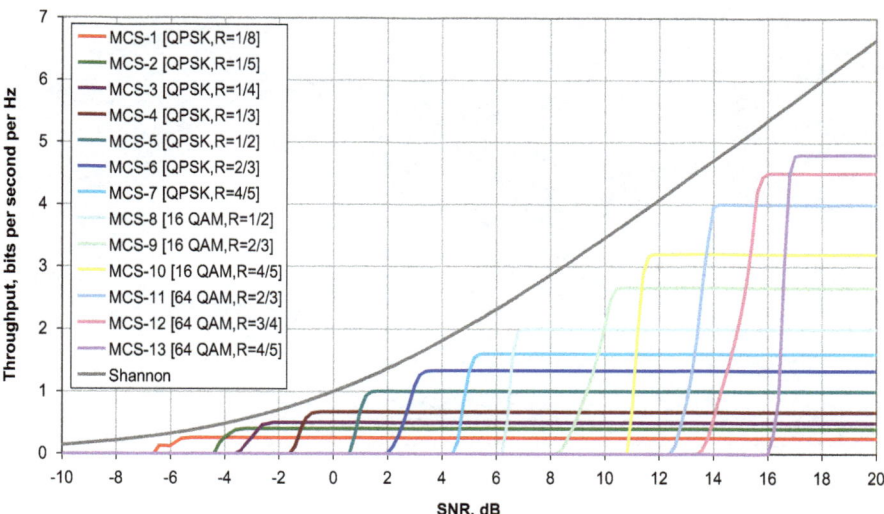

Fig. A.1 Throughput of a set of coding and modulation combinations, AWGN channels assumed (Wang et al., 2011; Wang, 2010; Wang et al., 2008)

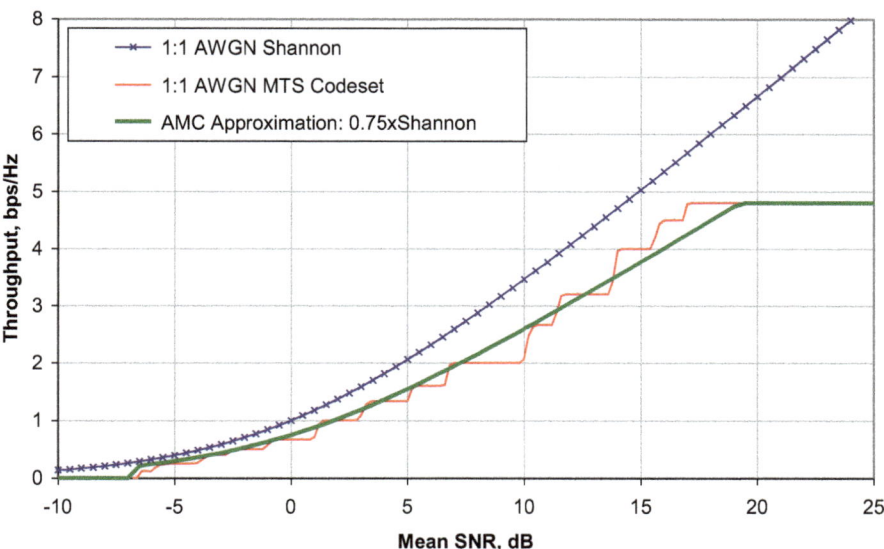

Fig. A.2 Approximating AMC with an attenuated and truncated form of the Shannon bound (Wang et al., 2011; Wang, 2010; Wang et al., 2008)

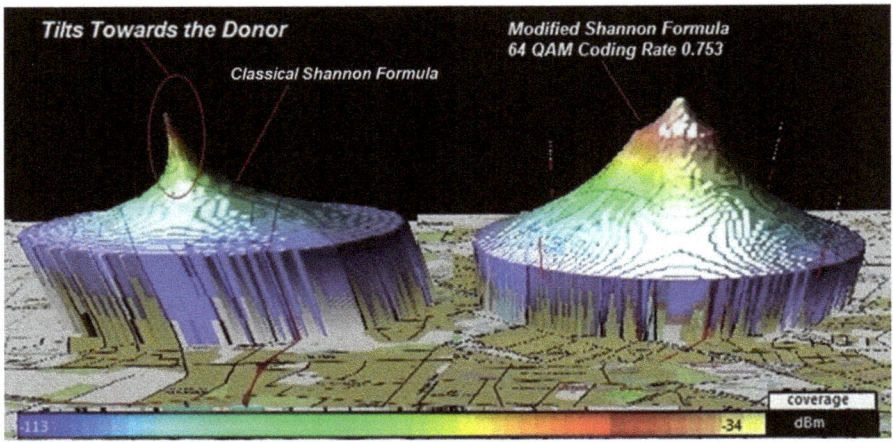

Fig. A.3 Performances of classical and modified Shannon capacity bound at three dimensions

References

Sesia, S., Toufik, I., & Baker, M. (2011). *LTE: the UMTS long term evolution*. Wiley.

Wang, F., Ghosh, A., Sankaran, C., Fleming, P., Hsieh, F., & Benes, S. (2008). Mobile WiMAX systems: performance and evolution. *Communications Magazine, IEEE, 46*(10), 41–49.

Wang, L.-C., Su, W.-S., Huang, J.-H., Chen, A., & Chang, C.-J. (2008). *Optimal relay location in multi-hop cellular systems*. Paper presented at the Wireless Communications and Networking Conference, 2008. WCNC 2008. IEEE.

Wang, Y. (2010). *System Level Analysis of LTE-Advanced: with Emphasis on Multi-Component Carrier Management*. Videnbasen for Aalborg UniversitetVBN, Aalborg UniversitetAalborg University, Det Teknisk-Naturvidenskabelige FakultetThe Faculty of Engineering and Science, Institut for Elektroniske SystemerDepartment of Electronic Systems.

Wang, J., Huang, Y., Zhong, C., Al-Qahtani, F., Wu, Q., & Cheng, Y. (2013). Performance analysis of interference-limited dual-hop multiple antenna AF relaying systems with feedback delay. *Eurasip Journal on Wireless Communications and Networking, 2013*(1), 1–15.

Appendix-B

Derivation of Eq. (3.83)

$$y_{RN}[t_1] = H_A X[t_1] + \sum_{q=1}^{Q} H_{B,q} X[t_1] + n_{RN} \tag{3.73}$$

$$y_{BS}[t_2] = \Psi H_A y_{RN}[t_1] + n_{BS} \tag{3.83}$$

By substitution of Eq. (3.73) in Eq. (3.83), the result is

$$y_{BS}[t_2] = \Psi H_A \left[H_A X[t_1] + \left(\sum_{q=1}^{Q} H_{B,q} X_q[t_1] \right) + n_{RN} \right] + n_{BS} \tag{3.84}$$

$$y_{BS}[t_2] = \left[\Psi H_A H_A X[t_1] + \Psi H_A \left(\sum_{q=1}^{Q} H_{B,q} X_q[t_1] \right) + \Psi H_A n_{RN} \right] + n_{BS} \tag{B.1}$$

$$y_{BS}[t_2] = \Psi H_A H_A X[t_1] + \Psi H_A \left(\sum_{q=1}^{Q} H_{B,q} X_q[t_1] \right) + \Psi H_A n_{RN} + n_{BS} \tag{B.2}$$

© Springer International Publishing Switzerland 2017

A. Yahya, *LTE-A Cellular Networks*, DOI 10.1007/978-3-319-43304-2

Appendix-C

Derivation of Eq. (3.86)

$$y_{\mathrm{RN}}[t_1] = H_A X[t_1] + \sum_{q=1}^{Q} H_{B,q} X[t_1] + n_{\mathrm{RN}} \qquad (3.73)$$

$$y_{\mathrm{UE},q}[t_2] = H_{C,q} X[t_2] + \Psi H_{B,q} y_{\mathrm{RN}}[t_1] + n_{\mathrm{UE}} \qquad (3.86)$$

By substitution of Eq. (3.73) in Eq. (3.86), the result is

$$y_{\mathrm{UE},q}[t_2] = H_{C,q} X[t_2] + \Psi H_{B,q} \left[H_A X[t_1] + \sum_{q=1}^{Q} H_{B,q} X[t_1] + n_{\mathrm{RN}} \right] + n_{\mathrm{UE}} \qquad (C.1)$$

$$
\begin{aligned}
y_{\mathrm{UE},q}[t_2] = {} & H_{C,q} X[t_2] + \Psi H_{B,q} \\
& \times \left[\Psi H_{B,q} H_A X[t_1] + \Psi H_{B,q} \sum_{q=1}^{Q} H_{B,q} X[t_1] + \Psi H_{B,q} n_{\mathrm{RN}} \right] + n_{\mathrm{UE}}
\end{aligned}
\qquad (C.2)
$$

$$
\begin{aligned}
y_{\mathrm{UE},q}[t_2] = {} & H_{C,q} X[t_2] + \Psi H_{B,q} H_A X[t_1] + \Psi H_{B,q} \sum_{q=1}^{Q} H_{B,q} X[t_1] \\
& + \Psi H_{B,q} n_{\mathrm{RN}} + n_{\mathrm{UE}}
\end{aligned}
\qquad (C.3)
$$

© Springer International Publishing Switzerland 2017
A. Yahya, *LTE-A Cellular Networks*, DOI 10.1007/978-3-319-43304-2

Appendix-D

Derivation of Amplification Factor (Ψ) for MR in Eq. (3.90)

$$\Psi = \sqrt{\frac{P_{\mathrm{RN}}}{|H_A|^2 P_i + |H_B|^2 P_{\mathrm{UE}} + N_O}} \tag{3.82}$$

$$\rho_{\mathrm{UE},q} = \frac{P_i |H_{C,q}|^2}{N_O} + \frac{\Psi^2 P_i |H_A|^2 |H_{B,q}|^2}{\left(\Psi^2 |H_{B,q}|^2 + 1\right) N_O} \tag{3.90}$$

Inserting Eq. (3.82) in Eq. (3.90), the instantaneous SNR at UE_q is

$$\rho_{\mathrm{UE},q} = \frac{P_i |H_{C,q}|^2}{N_O} + \frac{\left(\frac{P_{\mathrm{RN}}}{|H_A|^2 P_i + |H_B|^2 P_{\mathrm{UE}} + N_O}\right) P_i |H_A|^2 |H_{B,q}|^2}{\left(\left(\frac{P_{\mathrm{RN}}}{|H_A|^2 P_i + |H_B|^2 P_{\mathrm{UE}} + N_O}\right) |H_{B,q}|^2 + 1\right) N_O} \tag{D.1}$$

$$\rho_{\mathrm{UE},q} = \frac{P_i |H_{C,q}|^2}{N_O} + \frac{\left(\frac{P_{\mathrm{RN}} P_i |H_A|^2 |H_{B,q}|^2}{|H_A|^2 P_i + |H_B|^2 P_{\mathrm{UE}} + N_O}\right)}{\left(\left(\frac{P_{\mathrm{RN}} |H_{B,q}|^2}{|H_A|^2 P_i + |H_B|^2 P_{\mathrm{UE}} + N_O}\right) + 1\right) N_O} \tag{D.2}$$

$$\rho_{\mathrm{UE},q} = \frac{P_i |H_{C,q}|^2}{N_O} + \frac{\left(\frac{P_{\mathrm{RN}} P_i |H_A|^2 |H_{B,q}|^2}{|H_A|^2 P_i + |H_B|^2 P_{\mathrm{UE}} + N_O}\right)}{\left(\left(\frac{P_{\mathrm{RN}} |H_{B,q}|^2 + |H_A|^2 P_i + |H_B|^2 P_{\mathrm{UE}} + N_O}{|H_A|^2 P_i + |H_B|^2 P_{\mathrm{UE}} + N_O}\right)\right) N_O} \tag{D.3}$$

© Springer International Publishing Switzerland 2017
A. Yahya, *LTE-A Cellular Networks*, DOI 10.1007/978-3-319-43304-2

$$H_{k,q} = H_{B,q} = H_B \quad \text{(Same Channel)}$$
$$H_{c,q} = H_C \qquad\qquad \text{(Same Channel)} \qquad\qquad \text{(D.4)}$$
$$H_{A,q} = H_A \qquad\qquad \text{(Same Channel)}$$

$$\rho_{\text{UE},q} = \frac{P_i|H_{C,q}|^2}{N_\text{O}} + \frac{P_{\text{RN}}P_i|H_A|^2|H_{B,q}|^2}{(P_{\text{RN}}|H_{B,q}|^2 + |H_A|^2 P_i + |H_B|^2 P_{\text{UE}} + N_\text{O})N_\text{O}} \qquad \text{(D.5)}$$

Index

© Springer International Publishing Switzerland 2017
A. Yahya, *LTE-A Cellular Networks*, DOI 10.1007/978-3-319-43304-2

Lightning Source UK Ltd.
Milton Keynes UK
UKHW02f0015140918
328878UK00002B/7/P